Chemistry Fundamentals
IB

Second Edition

Pedro Patino
University of Central Florida

Kendall Hunt
publishing company

Kendall Hunt
publishing company

www.kendallhunt.com
Send all inquiries to:
4050 Westmark Drive
Dubuque, IA 52004-1840

Copyright © 2008, 2012 by Kendall Hunt Publishing Company

ISBN 978-1-4652-0064-8

Printed in the United States of America
10 9 8 7 6 5 4 3 2 1

CONTENTS

Chapter 7

Atomic Structure

Chapter Goals

- Describe the properties of electromagnetic radiation.
- Understand the origin of light from excited atoms and its relationship to atomic structure.
- Describe experimental evidence for wave-particle duality.
- Describe the basic ideas of quantum mechanics.
- Define the three quantum numbers (n, l, and m_l) and their relationship to atomic structure.

Electromagnetic Radiation

- light
- dual nature: wave and particle
- transverse wave: perpendicular oscillating electric and magnetic fields
- longitudinal wave: alternating areas of compression and decompression. <u>The direction of the wave is along the direction of propagation.</u>
- sound

Transverse Waves

- light
- do not require medium for propagation

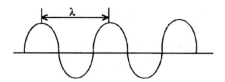

Amplitude

height of wave at maximum

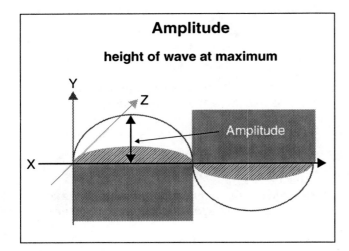

Wavelength, λ (lambda)

distance traveled by wave in 1 complete oscillation; distance from the top (crest) of one wave to the top of the next wave.

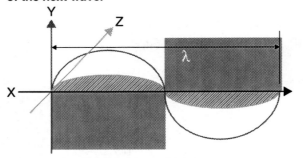

- λ measured in m, cm, nm, Å (angstrom)
- $1 \text{ Å} = 1 \times 10^{-10} \text{ m} = 1 \times 10^{-8} \text{ cm}$
- frequency, ν (**nu**), measured in s^{-1} (hertz) (Hz): number of **complete oscillations or cycles** passing a point per unit time (s)
- speed of propagation, distance traveled by ray per unit time in vacuum, all electromagnetic radiation travels at same rate

$c = 2.998 \times 10^{10}$ cm/s   (speed of light)

$\qquad = 2.998 \times 10^8$ m/s

$\qquad\qquad\qquad$ **(slower in air)**

$c\left(\frac{m}{s}\right) = \nu(s^{-1}) \times \lambda(m)$

What is the wavelength in nm of orange light, which has a frequency of 4.80×10^{14} s^{-1}?

$c = \lambda \times \nu$

$\lambda = \dfrac{c}{\nu} = \dfrac{2.998 \times 10^8 \text{ m s}^{-1}}{4.80 \times 10^{14} \text{ s}^{-1}} = 6.25 \times 10^{-7} \text{ m}$

$6.25 \times 10^{-7} \text{ m} \times \dfrac{1 \text{ nm}}{1 \times 10^{-9} \text{ m}} = 625 \text{ nm}$

Names to Remember

- Max Planck: quantized energy $E = h\nu$ ~ 1900
- Albert Einstein: photoelectric effect ~ 1905
- Niels Bohr: 2-D version of atom
- $E_n = (-R_H)(1/n^2)$ Balmer, 1885, then Bohr, 1913
- Louis de Broglie: wavelike properties of matter ~ 1915
- Werner Heisenberg: Uncertainty Principle ~ 1923
- Erwin Schrödinger: Schrödinger Equation ~ 1926

Planck's Equation

- Planck studied black-body radiation, such as that of a heated body, and realized that to explain the energy spectrum, he had to assume that:
  1. An object can gain or lose energy by absorbing or emitting radiant energy in **QUANTA** of specific frequency (ν)
  2. light has particle character (**photons**)

- Planck's equation is $\mathbf{E = h \times \nu = h \times \dfrac{c}{\lambda}}$
  E = energy of one photon

h = Planck's constant = 6.626×10^{-34} J.s/photon

Electromagnetic Spectrum

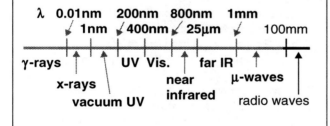

Compact disc players use lasers that emit red light with a wavelength of 685 nm. What is the energy of one photon of this light? What is the energy of one mole of photons of that red light?

λ, nm $\rightarrow \lambda$, m $\rightarrow \nu$, s$^{-1} \rightarrow$ E, J/photon $\rightarrow$ E, J/mole

$\times \dfrac{10^{-9}\,m}{nm}$ $\qquad \nu = \dfrac{c}{\lambda}$ $\qquad E = h\nu \times$ Avogadro's number

$$685 \text{ nm} \times \frac{10^{-9} \text{ m}}{1 \text{ nm}} = 6.85 \times 10^{-7} \text{ m}$$

$$\nu = \frac{c}{\lambda} = \frac{2.998 \times 10^8 \text{ m s}^{-1}}{6.85 \times 10^{-7} \text{ m}} = 4.38 \times 10^{14} \text{ s}^{-1}$$

$$E = h\nu = (6.626 \times 10^{-34} \text{ J·s/photon}) \times 4.38 \times 10^{14} \text{ s}^{-1}$$
$$= 2.90 \times 10^{-19} \text{ J/photon}$$

$$= (2.90 \times 10^{-19} \text{ J/photon}) \times 6.022 \times 10^{23} \text{ photons/mol}$$
$$= 1.75 \times 10^5 \text{ J/mol}$$

Example: Calculate the number of photons in a laser pulse with wavelength 337 nm and total energy 3.83 mJ.

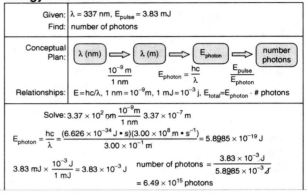

The Photoelectric Effect

- Light can strike the surface of some metals, causing electrons to be ejected.
- It demonstrates the particle nature of light.

The Photoelectric Effect

- What are some practical uses of the photoelectric effect?
- **Electronic door openers**
- **Light switches for street lights**
- **Exposure meters for cameras**
- Albert Einstein explained the effect
  – Explanation involved light having particle-like behavior.
  The minimum energy needed to eject the e⁻ is
  $E = h \times \nu$ (Planck's equation).
  It is also called "threshold" or binding energy.
  – Einstein won the 1921 Nobel Prize in Physics for this work.

Problem: An energy of 2.0×10^2 kJ/mol is required to cause a Cs atom on a metal surface to lose an electron. Calculate the longest possible λ of light that can ionize a Cs atom.

From the value of energy, we calculate the frequency (ν) and, with this we calculate lambda (λ).

Firstly, we need to calculate the energy in J per atom; it is given in kJ per mol of atoms:

$$2.0 \times 10^2 \, \frac{kJ}{mol} \times \frac{1000 \, J}{kJ} \times \frac{1 \, mol}{6.022 \times 10^{23} \, atoms} = 3.3 \times 10^{-19} \text{ Joule per atom}$$

$$E = h \times \nu \qquad \nu = \frac{E}{h} = \frac{3.3 \times 10^{-19} \, \text{Joule}}{6.626 \times 10^{-34} \, J \, s} = 5.0 \times 10^{14} \, s^{-1}$$

Now, speed of light, $c = \lambda \nu$ $\qquad \lambda = \frac{c}{\nu} = \frac{2.998 \times 10^8 \, m \, s^{-1}}{5.0 \times 10^{14} \, s^{-1}} = 6.0 \times 10^{-7} \, m$

$$6.0 \times 10^{-7} \, m \times \frac{1 \, nm}{1 \times 10^{-9} \, m} = 600 \, nm \qquad \text{(Visible light)}$$

Problem: A switch works by the photoelectric effect. The metal you wish to use for your device requires 6.7×10^{-19} J/atom to remove an electron. Will the switch work if the light falling on the metal has a $\lambda = 540$ nm or greater? Why?

The energy of photon is calculated with Planck's Equation.

$$E = h \times \nu = h \times \frac{c}{\lambda}$$
If calculated $E \geq 6.7 \times 10^{-19}$ J, the switch will work.

$$540 \, nm \times \frac{1 \times 10^{-9} \, m}{nm} = 5.40 \times 10^{-7} \, m$$

$$E = 6.626 \times 10^{-34} \, J \cdot s \times \frac{2.998 \times 10^8 \, m \, s^{-1}}{5.40 \times 10^{-7} \, m} = 3.68 \times 10^{-19} \, J$$

The switch won't open, because $E < 6.7 \times 10^{-19}$ J. λ has to be less than 540 nm.

Atomic Line Spectra and the Bohr Atom
(Niels Bohr, 1885 – 1962)

• An *emission spectrum* is formed by an electric current passing through a gas in a vacuum tube (at very low pressure), which causes the gas to emit light.
 – Sometimes called a *bright* line spectrum.

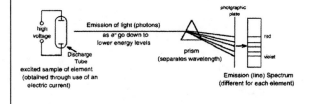

Atomic Line Spectra and the Bohr Atom

• **The Rydberg equation is an empirical equation that relates the wavelengths of the lines in the hydrogen spectrum. Lines are due to transitions:**

n_2 —— upper level

n_1 —— lower level

$$\frac{1}{\lambda} = R \left(\frac{1}{n_1^2} - \frac{1}{n_2^2} \right)$$

R is the Rydberg constant

$R = 1.097 \times 10^7 \, m^{-1}$

$n_1 < n_2$

N's refer to the numbers of the energy levels in the emission spectrum of hydrogen.

Example: What is the wavelength in angstroms of light emitted when the hydrogen atom's energy changes from n = 4 to n = 2?

$$n_2 = 4 \text{ and } n_1 = 2$$

$$\frac{1}{\lambda} = R\left(\frac{1}{n_1^2} - \frac{1}{n_2^2}\right)$$

$$\frac{1}{\lambda} = 1.097 \times 10^7 \, m^{-1}\left(\frac{1}{2^2} - \frac{1}{4^2}\right)$$

$$\frac{1}{\lambda} = 1.097 \times 10^7 \, m^{-1}(0.250 - 0.0625)$$

$$\frac{1}{\lambda} = 1.097 \times 10^7 \, m^{-1}(0.1875)$$

$$\frac{1}{\lambda} = 2.057 \times 10^6 \, m^{-1}$$

$$\lambda = \frac{1}{2.057 \times 10^6 \, m^{-1}} = 4.862 \times 10^{-7} \, m \times \frac{1 \, \text{Å}}{10^{-10} \, m} = 4862 \, \text{Å}$$

That corresponds to the green line in H spectrum.

The electron in a hydrogen atom relaxes from n = 7 with the emission of photons of light with a wave length of 397 nm. What is the final level of the electron after the transition?
The transition of the electron goes from $n_i = 7$ to $n_f = ?$

$$\underline{\hspace{3cm}} \, n_i = 7$$

$$\rightarrow \text{ photon } (E = h\nu) \quad \lambda = 397nm \times \frac{1 \times 10^{-9} \, m}{1 \, nm}$$

$$\underline{\hspace{2cm}} \, n_f = ?$$

$$= 3.97 \times 10^{-7} m$$

$$\frac{1}{\lambda} = 1.097 \times 10^7 m^{-1} \left(\frac{1}{n_f^2} - \frac{1}{7^2}\right)$$

CONTINUED: The electron in a hydrogen atom relaxes from n = 7 by emitting photons of light with a wave length of 397 nm. What is the final level of the electron after the transition?

$$\frac{1}{3.97 \times 10^{-7} m \times 1.097 \times 10^7 m^{-1}} = \frac{1}{n_f^2} - 2.041 \times 10^{-2}$$

$$\frac{1}{n_f^2} = 0.2296 + 0.02041$$

$$\frac{1}{n_f^2} = 0.25 \qquad n_f^2 = \frac{1}{0.25} = 4 \qquad n_f = 2$$

Atomic Line Spectra and the Bohr Atom

- An *absorption spectrum* is formed by shining a beam of white light through a sample of gas.
  - Absorption spectra indicate the wavelengths of light that have been <u>*absorbed*</u>.

- Every element has a unique spectrum.
- Thus, we can use spectra to identify elements.
- This can be done in the lab, stars, fireworks, etc.

Bohr Model of the Atom

- planetary model
- considers only the particle nature of the electron
- p^+ & n packed tightly in "**tiny**" nucleus
- electrons traveling in circular paths, orbits, in space surrounding nucleus
- size, energy, and e^- capacity of orbits increase as does distance from nucleus (orbital radius)
- orbits **quantized** (only certain levels exist)

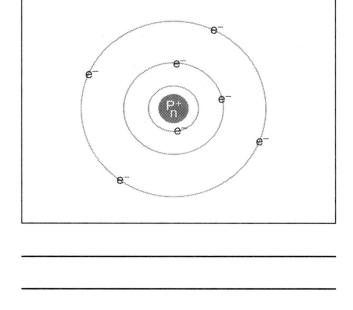

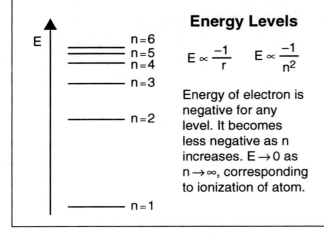

Energy Levels

$$E \propto \frac{-1}{r} \qquad E \propto \frac{-1}{n^2}$$

Energy of electron is negative for any level. It becomes less negative as n increases. $E \to 0$ as $n \to \infty$, corresponding to ionization of atom.

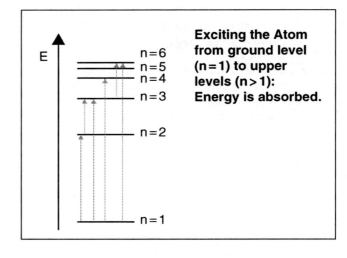

Exciting the Atom from ground level (n=1) to upper levels (n>1): Energy is absorbed.

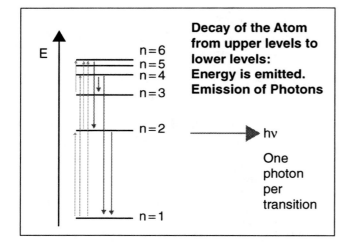

Decay of the Atom from upper levels to lower levels: Energy is emitted. Emission of Photons

$\longrightarrow$ hν

One photon per transition

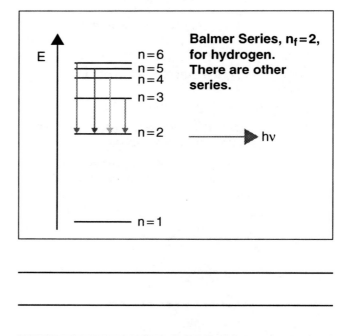

Balmer Series, $n_f=2$, for hydrogen. There are other series.

$\longrightarrow$ hν

Calculating E Difference Between Two Levels
A schoolteacher was the first to find this!
Johann Balmer

- $\Delta E = |E_{final} - E_{initial}| = R_H \left(\dfrac{1}{n_f{}^2} - \dfrac{1}{n_i{}^2} \right)$

- $R_H = 2.18 \times 10^{-18}$ J/atom = 1312 kJ/mol
- n_i and n_f = principal quantum numbers of the initial and final states: $n_f < n_i$
- 1, 2, 3, 4....

Problem: Calculate ΔE and λ for the violet line of Balmer series of H. $n_{initial} = 6$ $n_{final} = 2$

- $\Delta E = R_H \left(\dfrac{1}{n_f{}^2} - \dfrac{1}{n_i{}^2} \right)$ $R_H = 2.18 \times 10^{-18}$ J/atom

$\Delta E = 2.18 \times 10^{-18}$ J$\left(\dfrac{1}{2^2} - \dfrac{1}{6^2} \right) = 4.84 \times 10^{-19}$ J

$\Delta E = h\nu$ $\nu = c/\lambda$ Then, $\Delta E = hc/\lambda$

$\lambda = \dfrac{hc}{\Delta E} = \dfrac{6.626 \times 10^{-34} \text{ J.s} \times 2.998 \times 10^8 \text{ ms}^{-1}}{4.84 \times 10^{-19} \text{ J}}$

$\lambda = 4.104 \times 10^{-7}$ m $\times (1 \text{ Å}/10^{-10}) = 4104$ Å $= 410.4$ nm

Bohr Model of the Atom

- Bohr's theory correctly explains the H emission spectrum and those of hydrogen-like ions (He^+, Li^{2+} ... $1e^-$ species).

- The theory fails for atoms of all other elements because it is not an adequate theory: It doesn't take into account the fact that the (**very small**) electron can be thought as having wave behavior.

The Wave Nature of the Electron

- In 1925, Louis de Broglie published his PhD. dissertation.
  - A crucial element of his dissertation is that electrons have wave-like properties.
  - The electron wavelengths are described by the de Broglie relationship.

$\lambda = \dfrac{h}{mv}$

h = Planck's constant
m = mass of particle
v = velocity of particle

The Wave-Particle Duality of the Electron

- Consequently, we now know that electrons (in fact, all particles) have both a particle and a wave-like character.

- This wave-particle duality is a fundamental property of submicroscopic particles (**not for macroscopic ones.**)

The Wave-Particle Duality of the Electron

- Example: Determine the wavelength, in m and Å, of an electron, with mass 9.11×10^{-31} kg, having a velocity of 5.65×10^7 m/s.

$$h = 6.626 \times 10^{-34} \text{ Js} = 6.626 \times 10^{-34} \text{ kg m}^2/\text{s}$$

$$\lambda = \frac{h}{mv} = \frac{6.626 \times 10^{-34} \text{ kg m}^2\text{s}^{-1}}{9.11 \times 10^{-31}\text{kg} \times 5.65 \times 10^7 \text{ ms}^{-1}}$$

$$\lambda = 1.29 \times 10^{-11} \text{ m}$$

$$\lambda = 1.29 \times 10^{-11} \text{ m} \; \frac{1 \text{ Å}}{10^{-10} \text{ m}} = 0.129 \text{ Å}$$

Good: within atomic dimensions

The Wave-Particle Duality of the Electron

- Example: Determine the wavelength, in m, of a 0.22 caliber bullet, with mass 3.89×10^{-3} kg, having a velocity of 395 m/s, ~ 1300 ft/s.

$$h = 6.626 \times 10^{-34} \text{ Js} = 6.626 \times 10^{-34} \text{ kg m}^2/\text{s}$$

$$\lambda = \frac{h}{mv} = \frac{6.626 \times 10^{-34} \text{ kg m}^2\text{s}^{-1}}{3.89 \times 10^{-3} \text{ kg} \times 395 \text{ ms}^{-1}}$$

$$\lambda = 4.31 \times 10^{-34} \text{ m} = 4.31 \times 10^{-24} \text{ Å}$$

It's too small! It doesn't apply macro-objects!

Quantum Mechanical Model of the Atom
(considers both particle and wave nature of electrons)

- Heisenberg and Born in 1927 developed the concept of the **Uncertainty Principle:**
  It is impossible to determine simultaneously both the position and momentum of an electron (or any other small particle).

- Consequently, we must speak of the electrons' position about the atom in terms of **probability functions**, i.e., write a wave equation for each electron.

- These probability functions are represented as **orbitals** in quantum mechanics. They are the wave equations squared and plotted in 3 dimensions.

The Uncertainty Principle

Heisenberg showed that the more precisely the momentum (or the velocity) of a particle is known, the less precisely its position is known, and vice-versa:

$$(\Delta x)\, \Delta(mv) \geq \frac{h}{4\pi}$$

Δx: uncertainty in position

$\Delta(mv)$: uncertainty in momentum

h: Planck's constant

The Uncertainty Principle

Example: Determine the uncertainty in the position of an electron, with mass 9.11×10^{-31}kg, having a velocity of 5×10^6m/s. $h = 6.626\times10^{-34}$kg m^2/s

Assuming that uncertainty in velocity is 1%, $\Delta(mv) = m\,\Delta v$ and $\Delta v = (1/100)\, 5\times10^6$ m/s $= 5\times10^4$

$$\Delta x = \frac{h}{4\pi m\, \Delta v} = \frac{6.626\times10^{-34}\ kg\ m^2 s^{-1}}{4\pi\ 9.11\times10^{-31} kg \times 5\times10^4\ ms^{-1}}$$

$\Delta x = 1\times10^{-9}$ m $= 10$ Å!   **That means the uncertainty in the position is much bigger than the real size of an atom whose diameter is ~1–3 Å.**

We cannot say where the electron is!

Schrödinger's Model of the Atom
Basic Postulates of Quantum Theory

1. Atoms and molecules can exist only in certain energy states. In each energy state, the atom or molecule has a definite energy. When an atom or molecule changes its energy state, it must emit or absorb just enough energy to bring it to the new energy state (the quantum condition).

2. Atoms or molecules emit or absorb radiation (light) as they change their energies. The frequency of the light emitted or absorbed is related to the energy change by a simple equation.

$$E = h\, \nu = \frac{hc}{\lambda}$$

Schrödinger's Model of the Atom

3. The allowed energy states of atoms and molecules can be described by sets of numbers called quantum numbers.

- Quantum numbers are the solutions of the Schrödinger, Heisenberg & Dirac equations.

- **Four** quantum numbers are necessary to describe energy states of electrons in atoms.

Schrödinger equation

$$-\frac{b^2}{8\pi^2 m}\left(\frac{\partial^2\psi}{\partial^2 x} + \frac{\partial^2\psi}{\partial^2 y} + \frac{\partial^2\psi}{\partial^2 z} \right) + V\psi = E\psi$$

Orbital

- region of space within which one can expect to find an electron
- no solid boundaries
- electron capacity **of 2 per orbital**
- space surrounding nucleus divided up into large volumes called **shells**
- shells subdivided into smaller volumes called **subshells**
- orbitals located in subshells
- as shells get further from nucleus, energy, size, and electron capacity increase
- shells, subshells, and orbitals described by quantum numbers

Quantum Numbers

- The principal quantum number has the symbol n.
- $n = 1, 2, 3, 4, ...$ indicates **shell**
- K, L, M, N, ... **shells**
- As n increases, so does size, energy, and electron capacity.

  The electron's **energy depends** principally **on n.**

Quantum Numbers

- The angular momentum (**azimuthal**) quantum number has the symbol ℓ. It indicates subshell.

$\ell = 0, 1, 2, 3, 4, 5, ...(n-1)$
$\downarrow \downarrow \downarrow \downarrow \downarrow \downarrow$ from 0 to maximum $(n-1)$ for each n
$\ell = s, p, d, f, g, h, ...$ Subshells

- ℓ tells us the shape of the orbitals.
- These orbitals are the volume around the atom that the electrons occupy 90–95% of the time.

This is one of the places where Heisenberg's Uncertainty Principle comes into play.

Magnetic Quantum Number, m_l

- The symbol for the magnetic quantum number is m_ℓ. **It specifies the orientation of the orbital. For a given ℓ,**
$m_\ell = -\ell, (-\ell + 1), (-\ell + 2), ...0, ..., (\ell - 2), (\ell - 1), \ell$

- If $\ell = 0$ (or an s orbital), then $m_\ell = 0$.
  – Notice that there is only 1 value of m_ℓ.
    This implies that there is one s orbital per n value. $n \geq 1$

- If $\ell = 1$ (or a p orbital), then $m_\ell = -1, 0, +1$.
  – There are 3 values of m_ℓ.
    Thus, there are three p orbitals per n value. $n \geq 2$.

Magnetic Quantum Number, m_ℓ

- **If $\ell=2$** (or a d orbital), then $m_\ell=-2,-1,0,+1,+2$.
  - There are 5 values of m_ℓ.
  Thus there are five d orbitals per n value. $n \geq 3$

- **If $\ell=3$** (or an f orbital), then
  $m_\ell=-3,-2,-1,0,+1,+2,+3$.
  - There are 7 values of m_ℓ.   **$2\mathbf{x}\ell+1$ orbitals**
  Thus, there are seven f orbitals per n value, $n \geq 4$

- Theoretically, this series continues on to g, h, i, etc., orbitals.
  - Practically speaking atoms that have been discovered or made up to this point in time only have electrons in s, p, d, or f orbitals in their ground-state configurations.

n	shell	l	subshell	m_1	#orbitals	n^2	Max #e$^-$	
1	K	0	s	0	1	1	2	
2	L	0	s	0	1		2	
		1	p	$-1,0,1$	3	4	6	8
3	M	0	s	0	1		2	
		1	p	$-1,0,1$	3	9	6	18
		2	d	$-2,-1,0,1,2$	5		10	
4	N	0	s	0	1		2	
		1	p	$-1,0,1$	3		6	
		2	d	$-2,-1,0,1,2$	5	16	10	32
		3	f	$-3,-2,-1,0,1,2,3$	7		14	

Maximum two electrons per orbital

Electrons Indicated by Shell and Subshell

Symbolism

#electrons

$n\mathrm{l}^{\#}$

principal number   letter: s, p, d,.. orbital

$4s^1$ $3s^2$ $4d^{12}$ $5p^4$ $3p^7$ $4f^5$ $4f^{14}$

There are 4 electrons in the 5p orbitals.

The Shape of Atomic Orbitals

- s orbitals are spherically symmetric.

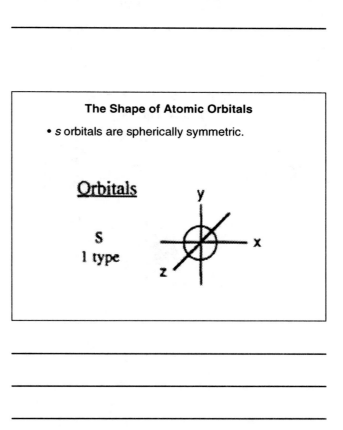

Orbitals

S
1 type

A plot of the surface density as a function of the distance from the nucleus for an *s* orbital of a hydrogen atom

• It gives the probability of finding the electron at a given distance from the nucleus.

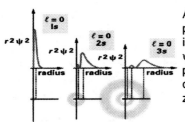

A node (nodal point or surface) is a point where ψ and the probability density (ψ^2) are zero.

---

p orbitals

• *p* orbital properties:
  – The first *p* orbitals appear in the n = 2 shell.

• *p* orbitals are peanut or dumbbell-shaped volumes.
  – They are directed along the axes of a Cartesian coordinate system.

• There are 3 *p* orbitals per n level.
  – The three orbitals are named p_x, p_y, p_z,
  – They have an $\ell = 1$.
  – $m_\ell = -1, 0, +1$ 3 values of m_ℓ

---

p Orbitals

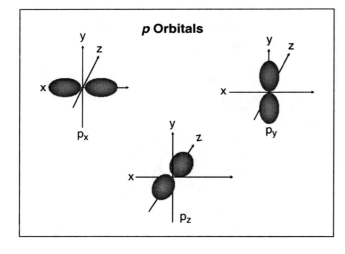

---

d orbitals

• *d* orbital properties:
  – The first *d* orbitals appear in the n = 3 shell.

• The five *d* orbitals have two different shapes:
  – 4 are cloverleaf shaped.
  – 1 is peanut shaped with a doughnut around it.
  – The orbitals lie directly on the Cartesian axes or are rotated 45° from the axes.

• There are 5 *d* orbitals per n level.
  – The five orbitals are named d_{xy}, d_{yz}, d_{xz}, $d_{x^2-y^2}$, d_{z^2}
  – They have an $\ell = 2$.
  – $m_\ell = -2, -1, 0, +1, +2$
    5 values of m_ℓ

d Orbitals

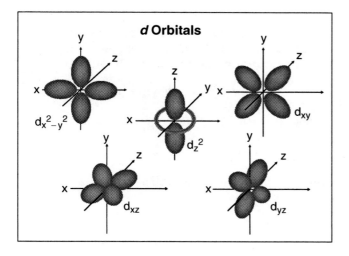

f orbitals

- f orbital properties:
  - <u>The first f orbitals appear in the n = 4 shell.</u>

- The f orbitals have the most complex shapes.

- There are seven f orbitals per n level.
  - The f orbitals have complicated names.
  - They have an $\ell = 3$
  - $m_\ell = -3, -2, -1, 0, +1, +2, +3$ 7 values of m_ℓ
  - The f orbitals have important effects in the lanthanide and actinide elements.

Problem: A possible excited state for the H atom has an electron in a 5d orbital. List all possible sets of quantum numbers n, l, and m_l for this electron.

$n = 5,$ $l = 2$ five possible m_l: $-2, -1, 0, +1, +2$
Then, there are five sets of   (n, l, m_l):

   (5, 2, -2)
   (5, 2, -1)
   (5, 2,   0)
   (5, 2, $+1$)
   (5, 2, $+2$)

Problem: Which of the following represent valid sets of quantum numbers? For a set that is invalid, explain briefly why it is not correct.

a) n = 3, l = 3, m_l: 0   No: maximum l = n−1

b) n = 2, l = 1, m_l: 0

c) n = 6, l = 5, m_l: −1

d) n = 4, l = 3, m_l: −4   No: minimum value of m_l: −l, that is m_l: −3

Problem: What is the maximum number of orbitals that can be identified by each of the following sets of quantum numbers? When "none" is the correct answer, explain the reason.

	Answer	Why?
a) $n = 4$, $l = 3$	Seven	m_l: $-3, -2, -1, 0, 1, 2, 3$
b) $n = 5$,	25	n^2
c) $n = 2$, $l = 2$	None	maximum $l = n - 1$
d) $n = 3$, $l = 1$, m_l: -1	One, just that described #s.	

_____

_____

_____

_____

Problem: State which of the following are incorrect designations for orbitals according to the quantum theory: 3p, 4s, 2f, and 1p

	Answer	Why?
a) 3p	correct	$n = 3$, $l = 1$ maximum $l = 3 - 1 = 2$
b) 4s	correct	
c) 2f	incorrect	maximum $l = 1$ ($l = 3$ for f)
d) 1p	incorrect	maximum $l = 0$ ($l = 1$ for p)

_____

_____

_____

_____

Chapter 8

Atomic Electron Configurations and Chemical Periodicity

Chapter Goals

- Understand the role magnetism plays in determining and revealing atomic structure.
- Understand effective nuclear charge and its role in determining atomic properties.
- Write the electron configuration of neutral atoms and monatomic ions.
- Understand the fundamental physical properties of the elements and their periodic trends.

Electron Spin and the Fourth Quantum Number

- The fourth quantum number is the spin quantum number, which has the symbol m_s.
- The spin quantum number has only two possible values.

    $m_s = +1/2$ or $-1/2$
    $m_s = \pm 1/2$

- This quantum number tells us the spin and orientation of the magnetic field of the electrons.
- Wolfgang Pauli discovered the Exclusion Principle in 1925.

No two electrons in an atom can have the same set of 4 quantum numbers, n, l, m_l, and m_s.

Electron Spin

- Spin quantum number effects:
  - Every orbital can hold up to two electrons.
    - Consequence of the Pauli Exclusion Principle.
  - The two electrons are designated as having:
    - one spin up $\uparrow$ $m_s = +1/2$
    - and one spin down $\downarrow$ $m_s = -1/2$
- Spin describes the direction of the electron's magnetic fields.

Paramagnetism and Diamagnetism

- Unpaired electrons have their spins alinged $\uparrow\uparrow$ or $\downarrow\downarrow$ **(in diff. orbitals)**
  - This increases the magnetic field of the atom.
  Total spin≠0, because they add up.

- Atoms with unpaired electrons are called *paramagnetic*.
  - Paramagnetic atoms are attracted to a magnet.

Paramagnetism and Diamagnetism

- Paired electrons have their spins unalinged $\uparrow\downarrow$. **(in the same orbital)**
  - Paired electrons have no net magnetic field.
  Total spin=0, because of cancellation.

- Atoms with no unpaired electrons are called *diamagnetic*.
  - Diamagnetic atoms are not attracted to a magnet.

Atomic Orbitals, Spin, and # of Electrons

- **Because two electrons in the same orbital must be paired (due to Pauli's Exclusion Principle), it is possible to calculate the number of orbitals and the number of electrons in each n shell.**
- The number of orbitals per n level is given by n^2 (see table at end of Chapter 7).
- The maximum number of electrons per n level is $2n^2$ (two electrons per orbital).
  - The value is $2n^2$ because of the two paired electrons per orbital.

n	shell	l	subshell	m_l	#orbitals	n^2	Max #e⁻	
1	K	0	s	0	1	1	2	
2	L	0	s	0	1	4	2	8
		1	p	−1,0,1	3		6	
3	M	0	s	0	1	9	2	18
		1	p	−1,0,1	3		6	
		2	d	−2,−1,0,1,2	5		10	
4	N	0	s	0	1	16	2	32
		1	p	−1,0,1	3		6	
		2	d	−2,−1,0,1,2	5		10	
		3	f	−3,−2,−1,0,1,2,3	7		14	

Atomic Subshell Energies and Electron Assignments

- The principle that describes how the periodic chart is a function of electronic configurations is the Aufbau Principle.
- The electron that distinguishes an element from the previous element enters the lowest energy atomic orbital available.

Atomic Subshell Energies and Electron Assignments

The Aufbau Principle describes the electron filling order in atoms. This is a product of the effective nuclear charge, Z*.
For the same n, Z* is higher for s orbital: $s > p > d > f$.
Then, e^- in s is the most attracted by nucleus and has the lowest energy.

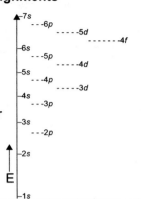

Atomic Subshell Energies and Electron Assignments

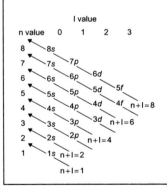

One mnemonic to remember the correct filling order for electrons in atoms is the increasing $n + \ell$ value.

Or, we can use this periodic chart?

	1A																		8A
P 1	1 H	2A												3A	4A	5A	6A	7A	2 He
2	3 Li	4 Be					(P–1) d							5 B	6 C	7 N	8 O	9 F	10 Ne
3	11 Na	12 Mg	3B	4B	5B	6B	7B	8B	8B	8B	1B	2B		13 Al	14 Si	15 P	16 S	17 Cl	18 Ar
4	19 K	20 Ca	21 Sc	22 Ti	23 V	24 Cr	25 Mn	26 Fe	27 Co	28 Ni	29 Cu	30 Zn		31 Ga	32 Ge	33 As	34 Se	35 Br	36 Kr
5	37 Rb	38 Sr	39 Y	40 Zr	41 Nb	42 Mo	43 Tc	44 Ru	45 Rh	46 Pd	47 Ag	48 Cd		49 In	50 Sn	51 Sb	52 Te	53 I	54 Xe
6	55 Cs	56 Ba	57 La	72 Hf	73 Ta	74 W	75 Re	76 Os	77 Ir	78 Pt	79 Au	80 Hg		81 Tl	82 Pb	83 Bi	84 Po	85 At	86 Rn
7	67 Fr	88 Ra	89 Ac	104 Rf	105 Db	106 Sg	107 Bh	108 Hs	109 Mt									(P) p	

(P) s

58 Ce	59 Pr	60 Nd	61 Pm	62 Sm	63 Eu	64 Gd	65 Tb	66 Dy	67 Ho	68 Er	69 Tm	70 Yb	71 Lu
90 Th	91 Pa	92 U	93 Np	94 Pu	95 Am	96 Cm	97 Bk	98 Cf	99 Es	100 Fm	101 Md	102 No	103 Lr

(P–2) f

Atomic Electron Configurations

- Now we will use the Aufbau Principle to determine the electronic configurations of the elements on the periodic chart.

- 1$\underline{st}$ row elements

	1s	Configuration
$_1$H	↑	$1s^1$
$_2$He	↑↓	$1s^2$

Atomic Electron Configurations

Hund's rule tells us that the electrons will fill the p and d orbitals by placing electrons in each orbital singly and with same spin until half-filled. That is the rule of maximum spin. Then the electrons will pair to finish the p orbitals.

	1s	2s	2p	Configuration
$_3$Li	↑↓	↑	_ _ _	$1s^2\,1s^1$
$_4$Be	↑↓	↑↓	_ _ _	$1s^2\,2s^2$
$_5$B	↑↓	↑↓	↑ _ _	$1s^2\,2s^2\,2p^1$
$_6$C	↑↓	↑↓	↑ ↑ _	$1s^2\,2s^2\,2p^2$
$_7$N	↑↓	↑↓	↑ ↑ ↑	$1s^2\,2s^2\,2p^3$
$_8$O	↑↓	↑↓	↑↓ ↑ ↑	$1s^2\,2s^2\,2p^4$
$_9$F	↑↓	↑↓	↑↓ ↑↓ ↑	$1s^2\,2s^2\,2p^5$
$_{10}$F	↑↓	↑↓	↑↓ ↑↓ ↑↓	$1s^2\,2s^2\,2p^6$

Electrons in orbitals of or same kind, such as p or d orbitals, in the same shell (n), have the same energy; they are said to be **degenerate.**

Atomic Electron Configurations

3rd row elements…

		3s	3p	Configuration
$_{11}$Na	[Ne]	↑	_ _ _	[Ne] $3s^1$
$_{12}$Mg	[Ne]	↑↓	_ _ _	[Ne] $3s^2$
$_{13}$Al	[Ne]	↑↓	↑ _ _	[Ne] $3s^2\,3p^1$
$_{14}$Si	[Ne]	↑↓	↑ ↑ _	[Ne] $3s^2\,3p^2$
$_{15}$P	[Ne]	↑↓	↑ ↑ ↑	[Ne] $3s^2\,3p^3$
$_{16}$S	[Ne]	↑↓	↑↓ ↑ ↑	[Ne] $3s^2\,3p^4$
$_{17}$Cl	[Ne]	↑↓	↑↓ ↑↓ ↑	[Ne] $3s^2\,3p^5$
$_{18}$Ar	[Ne]	↑↓	↑↓ ↑↓ ↑↓	[Ne] $3s^2\,3p^6$

Atomic Electron Configurations

4th row elements…

		3d	4s	4p	Configuration
$_{19}$K	[Ar]	_ _ _ _ _	↑	_ _ _	[Ar] $4s^1$

Atomic Electron Configurations

4th row elements…

	3d	4s	4p	Configuration
$_{19}$K [Ar]	_ _ _ _ _	↑	_ _ _	[Ar] $4s^1$
$_{20}$Ca [Ar]	_ _ _ _ _	↑↓	_ _ _	[Ar] $4s^2$

Atomic Electron Configurations

4th row elements…
The five **d** orbitals are **degenerate**.

	3d	4s	4p	Configuration
$_{19}$K [Ar]	_ _ _ _ _	↑	_ _ _	[Ar] $4s^1$
$_{20}$Ca [Ar]	_ _ _ _ _	↑↓	_ _ _	[Ar] $4s^2$
$_{21}$Sc [Ar]	↑ _ _ _ _	↑↓	_ _ _	[Ar] $4s^2\,3d^1$

Atomic Electron Configurations

4th row elements…
The five **d** orbitals are **degenerate**.

	3d	4s	4p	Configuration
$_{19}$K [Ar]	_ _ _ _ _	↑	_ _ _	[Ar] $4s^1$
$_{20}$Ca [Ar]	_ _ _ _ _	↑↓	_ _ _	[Ar] $4s^2$
$_{21}$Sc [Ar]	↑ _ _ _ _	↑↓	_ _ _	[Ar] $4s^2\,3d^1$
$_{22}$Ti [Ar]	↑ ↑ _ _ _	↑↓	_ _ _	[Ar] $4s^2\,3d^2$

Atomic Electron Configurations

4th row elements…
The five **d** orbitals are **degenerate**.

	3d	4s	4p	Configuration
$_{19}$K [Ar]	_ _ _ _ _	↑	_ _ _	[Ar] $4s^1$
$_{20}$Ca [Ar]	_ _ _ _ _	↑↓	_ _ _	[Ar] $4s^2$
$_{21}$Sc [Ar]	↑ _ _ _ _	↑↓	_ _ _	[Ar] $4s^2\,3d^1$
$_{22}$Ti [Ar]	↑ ↑ _ _ _	↑↓	_ _ _	[Ar] $4s^2\,3d^2$
$_{23}$V [Ar]	↑ ↑ ↑ _ _	↑↓	_ _ _	[Ar] $4s^2\,3d^3$

Atomic Electron Configurations

4th row elements… The [Ar] 4s^1 3d^5 configuration of **Cr** is more stable than [Ar] 4s^2 3d^4 **(expected)**.

	3d	4s	4p	Configuration
$_{19}$K [Ar]	_ _ _ _ _	↑	_ _ _	[Ar] 4s^1
$_{20}$Ca [Ar]	_ _ _ _ _	↑↓	_ _ _	[Ar] 4s^2
$_{21}$Sc [Ar]	↑ _ _ _ _	↑↓	_ _ _	[Ar] 4s^2 3d^1
$_{22}$Ti [Ar]	↑ ↑ _ _ _	↑↓	_ _ _	[Ar] 4s^2 3d^2
$_{23}$V [Ar]	↑ ↑ ↑ _ _	↑↓	_ _ _	[Ar] 4s^2 3d^3
$_{24}$Cr [Ar]	↑ ↑ ↑ ↑ ↑	↑	_ _ _	[Ar] 4s^1 3d^5

There is an extra measure of stability associated with half-filled and completely filled orbitals.

Atomic Electron Configurations

4th row elements… The [Ar] 4s^1 3d^{10} full d configuration of **Cu** is more stable than [Ar] 4s^2 3d^9 **(expected)**.

	3d	4s	4p	Configuration
$_{25}$Mn [Ar]	↑ ↑ ↑ ↑ ↑	↑↓	_ _ _	[Ar] 4s^2 3d^5
$_{26}$Fe [Ar]	↑↓ ↑ ↑ ↑ ↑	↑↓	_ _ _	[Ar] 4s^2 3d^6
$_{27}$Co [Ar]	↑↓ ↑↓ ↑ ↑ ↑	↑↓	_ _ _	[Ar] 4s^2 3d^7
$_{28}$Ni [Ar]	↑↓ ↑↓ ↑↓ ↑ ↑	↑↓	_ _ _	[Ar] 4s^2 3d^8
$_{29}$Cu [Ar]	↑↓ ↑↓ ↑↓ ↑↓ ↑↓ ↑	_ _ _		[Ar] 4s^1 3d^{10}

Another exception like Cr and for essentially the same reason.

Atomic Electron Configurations

4th row elements…

	3d	4s	4p	Configuration
$_{25}$Mn [Ar]	↑ ↑ ↑ ↑ ↑	↑↓	_ _ _	[Ar] 4s^2 3d^5
$_{26}$Fe [Ar]	↑↓ ↑ ↑ ↑ ↑	↑↓	_ _ _	[Ar] 4s^2 3d^6
$_{27}$Co [Ar]	↑↓ ↑↓ ↑ ↑ ↑	↑↓	_ _ _	[Ar] 4s^2 3d^7
$_{28}$Ni [Ar]	↑↓ ↑↓ ↑↓ ↑ ↑	↑↓	_ _ _	[Ar] 4s^2 3d^8
$_{29}$Cu [Ar]	↑↓ ↑↓ ↑↓ ↑↓ ↑↓ ↑	_ _ _		[Ar] 4s^1 3d^{10}
$_{30}$Zn [Ar]	↑↓ ↑↓ ↑↓ ↑↓ ↑↓	↑↓	_ _ _	[Ar] 4s^2 3d^{10}

Atomic Electron Configurations

4th row elements… … **(remember Hund's rule)**:
↑ ↑ _ is better (lower energy) than ↑↓ _ _
4p   4p

	3d	4s	4p	Configuration
$_{31}$Ga [Ar]	↑↓ ↑↓ ↑↓ ↑↓ ↑↓	↑↓	↑ _ _	[Ar] 4s^2 3d^{10} 4p^1
$_{32}$Ge [Ar]	↑↓ ↑↓ ↑↓ ↑↓ ↑↓	↑↓	↑ ↑ _	[Ar] 4s^2 3d^{10} 4p^2
$_{33}$As [Ar]	↑↓ ↑↓ ↑↓ ↑↓ ↑↓	↑↓	↑ ↑ ↑	[Ar] 4s^2 3d^{10} 4p^3
$_{34}$Se [Ar]	↑↓ ↑↓ ↑↓ ↑↓ ↑↓	↑↓	↑↓ ↑ ↑	[Ar] 4s^2 3d^{10} 4p^4
$_{35}$Br [Ar]	↑↓ ↑↓ ↑↓ ↑↓ ↑↓	↑↓	↑↓ ↑↓ ↑	[Ar] 4s^2 3d^{10} 4p^5
$_{36}$kr [Ar]	↑↓ ↑↓ ↑↓ ↑↓ ↑↓	↑↓	↑↓ ↑↓ ↑↓	[Ar] 4s^2 3d^{10} 4p^6

Atomic Electron Configurations

Lanthanides (4f)

$_{56}$Ba [Xe] $6s^2$

$_{57}$La [Xe] $5d^1 6s^2$

$_{58}$Ce [Xe] $4f^1 5d^1 6s^2$

$_{59}$Pr [Xe] $4f^3 6s^2$ Praseodymium

$_{70}$Yb [Xe] $4f^{14} 6s^2$ Ytterbium

$_{71}$Lu [Xe] $4f^{14} 5d^1 6s^2$ Lutetium

---

s, p, d, and f-block in the Periodic Table

P	1A																	8A
1	1 H	2A										3A	4A	5A	6A	7A		2 He
2	3 Li	4 Be					(P−1) d						5 B	6 C	7 N	8 O	9 F	10 Ne
3	11 Na	12 Mg	3B	4B	5B	6B	7B	8B	8B	8B	1B	2B	13 Al	14 Si	15 P	16 S	17 Cl	18 Ar
4	19 K	20 Ca	21 Sc	22 Ti	23 V	24 Cr	25 Mn	26 Fe	27 Co	28 Ni	29 Cu	30 Zn	31 Ga	32 Ge	33 As	34 Se	35 Br	36 Kr
5	37 Rb	38 Sr	39 Y	40 Zr	41 Nb	42 Mo	43 Tc	44 Ru	45 Rh	46 Pd	47 Ag	48 Cd	49 In	50 Sn	51 Sb	52 Te	53 I	54 Xe
6	55 Cs	56 Ba	57 La	72 Hf	73 Ta	74 W	75 Re	76 Os	77 Ir	78 Pt	79 Au	80 Hg	81 Tl	82 Pb	83 Bi	84 Po	85 At	86 Rn
7	87 Fr	88 Ra	89 Ac	104 Rf	105 Db	106 Sg	107 Bh	108 Hs	109 Mt								(P) p	

(P) s

(P−2) f	58 Ce	59 Pr	60 Nd	61 Pm	62 Sm	63 Eu	64 Gd	65 Tb	66 Dy	67 Ho	68 Er	69 Tm	70 Yb	71 Lu
	90 Th	91 Pa	92 U	93 Np	94 Pu	95 Am	96 Cm	97 Bk	98 Cf	99 Es	100 Fm	101 Md	102 No	103 Lr

---

Valence Electrons

Electrons in shell with highest n, i.e., **the outermost electrons**, those beyond the core electrons

$1s^2\ 2s^2\ 2p^6\ \mathbf{3s^1}$

$1s^2\ 2s^2\ 2p^6\ \mathbf{3s^2\ 3p^2}$

$1s^2\ 2s^2\ 2p^6\ 3s^2\ 3p^6\ 3d^{10}\ \mathbf{4s^2\ 4p^6}$

$1s^2\ \mathbf{2s^2}$

$1s^2\ 2s^2\ 2p^6\ 3s^2\ 3p^6\ \mathbf{4s^2\ 3d^7}$

They determine the chemical properties of an element. For the representative elements, they are the **ns** and **np** electrons; for transition elements, they are the ns and (n−1)d electrons.

P	1A	
1	1 H	$1s^1$
2	3 Li	$2s^1$
3	11 Na	$3s^1$
4	19 K	$4s^1$ # of valence electrons = 1
5	37 Rb	$5s^1$
6	55 Cs	$6s^1$
7	87 Fr	$7s^1$

Slide 1

2A

4 Be	$2s^2$
12 Mg	$3s^2$
20 Ca	$4s^2$
38 Sr	$5s^2$
56 Ba	$6s^2$
88 Ra	$7s^2$

# of valence electrons = 2

Slide 2

3A

5 B	$2s^2 2p^1$
13 Al	$3s^2 3p^1$
31 Ga	$4s^2 4p^1$
49 In	$5s^2 5p^1$
81 Tl	$6s^2 6p^1$

# of valence electrons = 3

Slide 3

of valence electrons = 7

7A

9 F	$2s^2 2p^5$
17 Cl	$3s^2 3p^5$
35 Br	$4s^2 4p^5$
53 I	$5s^2 5p^5$
85 At	$6s^2 6p^5$

For the representative elements, the
# of valence electrons = # of group.

Slide 4

The element X has the valence shell
electron configuration, $ns^2 np^4$. X
belongs to what group?
chalcogens

1A																		8A
1 H	2A											3A	4A	5A	6A	7A		2 He
3 Li	4 Be											5 B	6 C	7 N	8 O	9 F		10 Ne
11 Na	12 Mg	3B	4B	5B	6B	7B	8B	8B	8B	1B	2B	13 Al	14 Si	15 P	16 S	17 Cl		18 Ar
19 K	20 Ca	21 Sc	22 Ti	23 V	24 Cr	25 Mn	26 Fe	27 Co	28 Ni	29 Cu	30 Zn	31 Ga	32 Ge	33 As	34 Se	35 Br		36 Kr
37 Rb	38 Sr	39 Y	40 Zr	41 Nb	42 Mo	43 Tc	44 Ru	45 Rh	46 Pd	47 Ag	48 Cd	49 In	50 Sn	51 Sb	52 Te	53 I		54 Xe
55 Cs	56 Ba	57 La	72 Hf	73 Ta	74 W	75 Re	76 Os	77 Ir	78 Pt	79 Au	80 Hg	81 Tl	82 Pb	83 Bi	84 Po	85 At		86 Rn
87 Fr	88 Ra	89 Ac	104 Unq	105 Unp	106 Unh	107 Uns	108 Uno	109 Une										

Energy (Orbital) Diagram

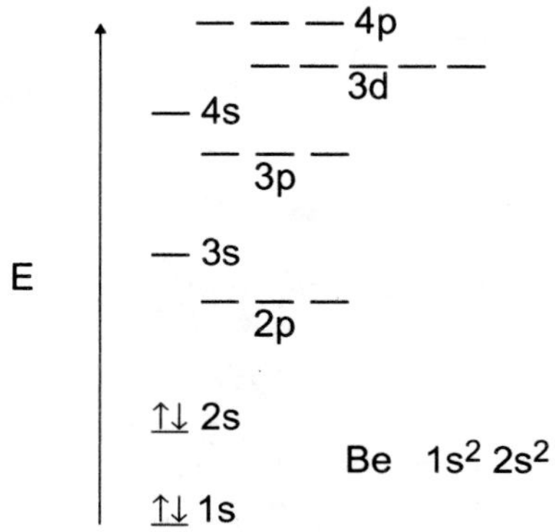

$$— — —\ 4p$$
$$— —\ \overline{3d}\ — —$$
$$—\ 4s$$
$$—\ \overline{3p}\ —$$
$$—\ 3s$$
$$—\ \overline{2p}\ —$$

E

$\uparrow\downarrow$ 2s

Be $1s^2\ 2s^2$

$\uparrow\downarrow$ 1s

Orbital Box Diagrams

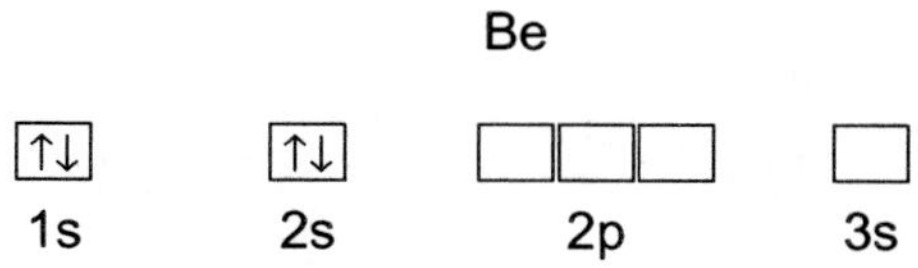

Be

$\uparrow\downarrow$	$\uparrow\downarrow$				
1s	2s		2p		3s

Orbital Box Diagrams

N

$\uparrow\downarrow$	$\uparrow\downarrow$	$\uparrow$	$\uparrow$	$\uparrow$
1s	2s		2p	

Formation of Cations

electrons lost from subshell with highest
n and l first (**from valence electrons**)

examples

K $1s^2\ 2s^2\ 2p^6\ 3s^2\ 3p^6\ 4s^1$
[Ar] **$4s^1$**

K$^+$ $1s^2\ 2s^2\ 2p^6\ 3s^2\ 3p^6$
[Ar]

Ca $1s^2\ 2s^2\ 2p^6\ 3s^2\ 3p^6\ 4s^2$
   [Ar] **$4s^2$**

Ca^{2+} $1s^2\ 2s^2\ 2p^6\ 3s^2\ 3p^6$
   [Ar]

Al $1s^2\ 2s^2\ 2p^6\ \mathbf{3s^2\ 3p^1}$
Al^{3+} [Ne]

In [Kr] $4d^{10}\ \mathbf{5s^2\ 5p^1}$
In^{3+} [Kr] $4d^{10}$

Transition Metal Cations

In the process of ionization of the transition metals, the ns electrons are lost before the (n-1)d:

Fe: [Ar] $3d^6\ 4s^2$ → Fe^{2+}: [Ar] $3d^6$

Fe^{2+}: [Ar] $3d^6$ → Fe^{3+}: [Ar] $3d^5$

Cu: [Ar] $3d^{10}\ 4s^1$ → Cu^+: [Ar] $3d^{10}$

Cu^+: [Ar] $3d^{10}$ → Cu^{2+}: [Ar] $3d^9$

Fe, Fe^{2+}, Fe^{3+}, Cu, and Cu^{2+} are paramagnetic

Two problems of ions, charge, and electron configuration

An anion has a 3– charge and electron configuration $1s^2\ 2s^2\ 2p^6\ 3s^2\ 3p^6$. What is the symbol of the ion?
  The neutral atom has gained 3e– to form the ion, then the neutral atom had 15 e–. In the neutral atom, the # e– = # p^+ = Atomic number, that is 15. The element is, then, phosphorus (phosphorus). Symbol of ion is P^{3-}.

A cation has a 2^+ charge, and its electron configuration is [Ar] $3d^7$. What is the symbol of the ion?
  Here, the neutral atom has lost 2e–. It is a transition metal, due to the 3d electrons. Remember, they first lose e–s in 4s orbital. Symbol of ion is Co^{2+}.
Neutral atom has $18 + 7 + 2 = 27$ e¨ = 27 p^+ = atomic #
   [Ar] $3d^7$ lost

Atomic Properties and Periodic Trends

Periodic Properties of

the Elements
1. Atomic Radii
2. Ionization Energy
3. Electron Affinity
4. Ionic Radii

Atomic Properties and Periodic Trends

- Establish a classification scheme of the elements based on their electron configurations.
- **Noble Gases**
  - All of them have completely filled electron shells. They are not very reactive.
- Since they have similar electronic structures, their chemical reactions are similar.
  - He $1s^2$
  - Ne   [He] $2s^2 2p^6$
  - Ar   [Ne] $3s^2 3p^6$
  - Kr   [Ar] $4s^2 4p^6$
  - Xe   [Kr] $5s^2 5p^6$
  - Rn   [Xe] $6s^2 6p^6$

Atomic Properties and Periodic Trends

Representative Elements are the elements in A groups on periodic chart.

These elements will have their "last" electron in an outer *s* or *p* orbital.

These elements have fairly regular variations in their properties.

Metallic character, for example, increases from right to left and top to bottom.

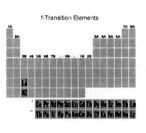
Representative Elements

Atomic Properties and Periodic Trends

- ***d*-Transition Elements**
  Elements on periodic chart in B groups. Sometimes called transition metals.
- Each metal has *d* electrons.
  n*s* (n–1)*d* configurations
- These elements make the transition from metals to nonmetals.
- Exhibit smaller variations from row-to-row than the representative elements.

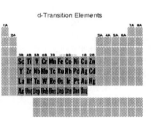
d-Transition Elements

Atomic Properties and Periodic Trends

- ***f*-transition metals**
  Sometimes called inner transition metals.
- Electrons are being added to *f* orbitals.
- Electrons are being added two shells below the valence shell!
- Consequently, very slight variations of properties from one element to another.

f-Transition Elements

Atomic Properties and Periodic Trends

**Outermost** electrons (valence electrons) have the greatest Influence on the chemical properties of elements.

Atomic Properties and Periodic Trends

Atomic radii describe the relative sizes of atoms.

Atomic radii increase within a column going from the top to the bottom of the periodic table.

The outermost electrons are assigned to orbitals with increasingly higher values of n.

The underlying electrons require some space, so the electrons of the outer shells must be further from the nucleus.

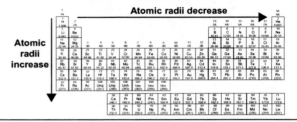

Atomic Properties and Periodic Trends

Atomic radii decrease within a row going from left to right on the periodic table.

This last fact seems contrary to intuition.

How does nature make the elements smaller even though the electron number is increasing?

Atomic Radii

- The reason the atomic radii decrease across a period is due to **shielding** or **screening** effect.
  - Effective nuclear charge, Z_{eff}, experienced by an electron is less than the actual nuclear charge, Z.
  - The inner electrons block the nuclear charge's effect on the outer electrons.
- Moving across a period, each element has an increased nuclear charge and the electrons are going into the same shell (2s and 2p or 3s and 3p, etc.).
  - Consequently, the outer electrons feel a stronger effective nuclear charge.
  - For Li, $Z_{eff} \sim +1$
  - For Be, $Z_{eff} \sim +2$

Atomic Radii

- Example: Arrange these elements based on their **increasing** atomic radii.
  - Se, S, O, Te

  O < S < Se < Te

  In the same group, atomic size increases as n (and Z) increases.

  - Br, Ca, Ge, F

       F < Br < Ge < Ca

  same group   same period

Ionization Energy

- **First ionization energy (IE$_1$)**
  - **The minimum amount of energy required to remove the most loosely bound electron from an isolated gaseous atom to form a 1+ ion.**
- **Symbolically:**
  Atom$_{(g)}$ + energy   ion$^+_{(g)}$ + e$^-$

$Mg_{(g)}$ + 738 kJ/mol $\rightarrow$ Mg^+ + e^- IE$_1$ = 738 kJ/mol

$1s^2\,2s^2\,2p^6\,3s^2$ $1s^2\,2s^2\,2p^6\,3s^1$

Ionization Energy

- **Second ionization energy (IE$_2$)**
  - **The amount of energy required to remove the second electron from a gaseous 1+ ion.**
- **Symbolically:**
  - **ion$^+$ + energy $\rightarrow$ ion^{2+} + e$^-$**

Mg^+ + 1451 kJ/mol $\rightarrow$ $Mg2^+$ + e^- IE$_2$ = 1451 kJ/mol

$1s^2\,2s^2\,2p^6\,3s^2$ $1s^2\,2s^2\,2p^6\,3s^1$

Atoms can have 3<u>rd</u> (IE$_3$), 4<u>th</u> (IE$_4$), etc. ionization energies. The values are getting larger.

Ionization Energy

Periodic trends for Ionization Energy:
1) IE$_2$ > IE$_1$: It always takes more energy to remove a second electron from an ion than from a neutral atom.
2) IE$_1$ generally increases moving from IA elements to VIIIA elements.
Important exceptions at Be & Mg, N & P, etc., due to filled and half-filled subshells.
3) IE$_1$ generally decreases moving down a family.
IE$_1$ for Li > IE$_1$ for Na, etc

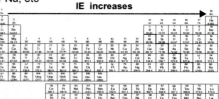

Ionization Energy

- Example: Arrange these elements based on their first ionization energies.
  - Sr, Be, Ca, Mg

    Sr < Ca < Mg < Be

  - Al, Cl, Na, P

    Na < Al < P < Cl

  - O, Ga, Sr, Se

    Sr < Ga < Se < O

Ionization Energy

- The reason Na forms Na^+ and not Na^{2+} is that the energy difference between IE_1 and IE_2 is so large.
  - Requires more than 9 times more energy to remove the second electron than the first one.

- The same trend is persistent throughout the series.
  - Thus, Mg forms Mg^{2+} and not Mg^{3+}.
  - Al forms Al^{3+} and not Al^{4+}

Ionization Energies (kJ/mole)

	1312							
He	2371	5247						
Li	520	7297	11810					
Be	900	1757	14840	21000				
B	800	2430	3659	25020	32810			
C	1086	2352	4619	6221	37800	47300		
N	1402	2857	4577	7473	9443	53250	64340	
O	1314	3391	5301	7468	10980	13320	71300	84850
F	1681	3375	6045	8418	11020	15160	17860	92000
Ne	2080	3963	6276	9376	12190	15230		
Na	496	4565	6912	9540	13360	16610	20110	25490
Mg	738	1450	7732	10550	13620	18000	21700	25660
Al	577	1816	2744	11580	15030	18370	23290	27460
Si	786	1577	3229	4356	16080	19790	23780	29250
P	1012	1896	2910	4954	6272	21270	25410	29840
S	1000	2260	3380	4565	6996	8490	28080	31720
Cl	1255	2297	3850	5146	6544	9330	11020	33600
Ar	1520	2665	3947	5770	7240	8810	11970	13840
K	419	3069	4600	5879	7971	9619	11380	14950

Electron Affinity (EA)

- Electron affinity is the amount of energy **absorbed or emitted** when an electron is added to an isolated gaseous atom to form an ion with a 1- charge.
- Sign conventions for electron affinity.
  - If EA > 0, energy is absorbed (difficult)
  - If EA < 0, energy is released (easy)
- Electron affinity is a measure of an atom's ability to form negative ions.
- Symbolically:

  $atom(g) + e^- \rightarrow ion^-(g)$ **EA (kJ/mol)**

Electron Affinity

- General periodic trends for electron affinity are:
  - the values become more negative from left to right across a period on the periodic chart **(affinity for electron increases).**
  - the values become more negative from bottom to top, at a group on the periodic chart.
- **Noble gases have EA > 0 (full electron confg)**
- An element with a high ionization energy generally has a high affinity for an electron, i.e., EA is largely negative. That is the case for halogens (F, Cl, Br, I), O, and S.

Electron Affinity

F (Z = 9) and Cl(Z = 17) have the most negative EA

For noble gases, He (2), Ne (10), and Ar(18), EA > 0

Electron Affinity

Two examples of electron affinity values:

$Mg(g) + e^- + 231\ kJ/mol \rightarrow Mg^-(g)$ $EA = 231\ kJ/mol$

$Br(g) + e^- \rightarrow Br^-(g) + 323\ kJ/mol$ $EA = -323\ kJ/mol$

Br has a larger affinity for e^- than Mg. The greater the affinity an atom has for an e^-, the more negative EA is.

Affinity for electrons increases. EA decreases, it becomes more negative.

Affinity for electrons decreases.

EA value increases; it becomes less negative.

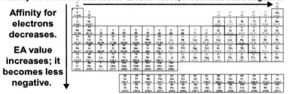

Ionic Radii

⌗Cations (positive ions) are always *smaller* than their respective neutral atoms. When one or more electrons are removed, the attractive force of the protons is now exerted on less electrons.

Element	Na 11 p+, 11 e-	Mg 12 p+, 12 e-	Al 13 p+, 13 e-
Atomic Radius (Å)	1.86	1.60	1.43
Ion	Na+ 11 p+, 10 e-	Mg2+ 12 p+, 10 e-	Al3+ 13 p+, 10 e-
Ionic Radius (Å)	1.16	0.85	0.68

Ionic Radii

#Anions (negative ions) are always *larger* than their neutral atoms.

$$F\ 1s^2\ 2s^2\ 2p^5 + e^- \rightarrow F^-1s^2\ 2s^2\ 2p^6 \qquad \text{same Z}$$

nine electrons   ten electrons

Element	N $7\ p^+,\ 7\ e^-$	O	F
Atomic Radius (Å)	0.75	0.73	0.72
Ion	N^{3-} $7\ p^+,\ 10\ e^-$	O^{2-}	F^-
Ionic Radius (Å)	1.71	1.26	1.19

Ionic Radii

Cation (positive ions) radii decrease from left to right across a period.

Increasing nuclear charge attracts the electrons and decreases the radius.

Rb^+ and Sr^{2+} are <u>isoelectronic</u>, same # of e-s

Ion	Rb^+	Sr^{2+}	In^{3+}
Ionic Radii (Å)	1.66	1.32	0.94

Ionic Radii

Anion (negative ions) radii decrease from left to right across a period.

Increasing electron numbers in highly charged ions cause the electrons to repel and increase the ionic radius.

For these **isoelectronic** anions…

10 e⁻ and 7 p⁺   8 p⁺   9 p⁺

Ion	N^{3-}	O^{2-}	F^-
Ionic Radii(Å)	1.71	1.26	1.19

Ionic Radii

Example: Arrange these ions in order of decreasing radius.

Ga^{3+}, K^+, Ca^{2+}
$K^+ > Ca^{2+} > Ga^{3+}$

Cl^-, Se^{2-}, Br^-, S^{2-}
$Se^{2-} > Br^- > S^{2-} > Cl^-$

isoelectronic   isoelectronic

$Se^{2-}(34\ p^+) > Br^-(35\ p^+)$; they have 36 e⁻ each.

$S^{2-}(16\ p^+) > Cl^-(17\ p^+)$; they have 18 e⁻ each.

$Br^- > S^{2-}$ because Br^- is in the 4th period, S^{2-} is in the 3rd.

Ionic Radii of Isoelectronic Species

Isoelectronic species have the same number of electrons. Here are some examples with the number of (protons) and + or − charges.

$N^{3-}(Z=7) > O^{2-}(Z=7) > F^-(Z=7) > Ne(Z=10)$ **neutral** >

$Na^+(Z=11) > Mg^{2+}(Z=12) > Al^{3+}(Z=13)$ **all have 10e⁻**

The nuclear charge (+) increases from left to right, so do attraction force to electrons: r decreases.

$S^{2-}(Z=16) > Cl^-(Z=17) > Ar(Z=18) > K^+(Z=19) >$

$Ca^{2+}(Z=20) > Sc^{3+}(Z=21)$ **all have 18e⁻**

_____

_____

_____

_____

Chapter 9

Chemical Bonding

Chapter Goals

- Understand the difference between ionic and covalent bonds.
- Draw Lewis electron dot structures for small molecules and ions.
- Use the valence shell electron-pair repulsion theory (VSEPR) to predict the shapes of simple molecules and ions and to understand the structure of more complex molecules.
- Use electronegativity to predict the charge distribution in molecules and ions and to define the polarity of bonds.
- Predict the polarity of molecules.
- Understand the properties of covalent bonds and their influence on molecular structure.

Introduction: Valence Electrons

- Attractive forces that hold atoms together in compounds are called chemical bonds.
- The electrons involved in bonding are usually those in the outermost (valence) shell. **The (inner) core electrons are not involved in chemical behavior.**

Valence electrons:
- For the main group's (representative) elements, they are the outer most **s** and **p** electrons. The # of valence electrons is equal to the group number.
- For the transition elements, they are the **ns** and **(n–1)d** electrons.

Lewis Dot Symbols for Atoms

- Lewis dot formulas or Lewis dot symbols are a convenient bookkeeping method for tracking **valence electrons.**
- Valence electrons are those electrons that are transferred or involved in chemical bonding. **They are chemically important.**

Symbol considered to have 4 sides, two dots per side:

Ḣ   Ḧe

L̇i   B̈e   Ḃ·   ·Ċ·   ·N̈·   :Ö·   :F̈·   :N̈e:

Lewis Dot Symbols for Atoms

Elements that are in the same periodic group have the same Lewis dot structures.

Li & Na   ·N· & ·P·   :F· & :Cl·

Formation of Bonds

When a chemical reaction occurs, the valence electrons of the atoms are reorganized so that net attractive forces—chemical bonds—occur between atoms.

Chemical bonds are classified into two types:

- *Ionic bonding* results from electrostatic attractions between ions, which are formed by the <u>transfer</u> of one or more electrons from one atom to another.

- *Covalent bonding* results from <u>sharing</u> one or more electron pairs between two atoms.

Comparison of Ionic and Covalent Compounds

<u>Melting point comparison</u>
Ionic compounds are usually solids with high melting points.

    Typically > 400°C

Covalent compounds are gases, liquids, or solids with low melting points.

    Typically < 300°C

<u>Solubility in polar solvents (such as water)</u>
Ionic compounds are generally soluble.
Covalent compounds are generally insoluble.

Comparison of Ionic and Covalent Compounds

<u>Solubility in nonpolar solvents</u>
Ionic compounds are generally insoluble.
Covalent compounds are generally soluble.

<u>Conductivity in molten solids and liquids</u>
Ionic compounds generally conduct electricity. They contain mobile ions.
Covalent compounds generally do not conduct electricity.

Comparison of Ionic and Covalent Compounds

Conductivity in aqueous solutions
  Ionic compounds generally conduct electricity. They contain mobile ions.
  Covalent compounds are poor conductors of electricity.

Formation of Compounds
  Ionic compounds are formed between elements with large differences in **electronegativity.**
    Often a metal and a nonmetal
  Covalent compounds are formed between elements with similar **electronegativities.**
    Usually two or more nonmetals

Ionic Bonding

Formation of Ionic Compounds

An **ion** is an atom or a group of atoms possessing a net electrical charge.

Ions come in two basic types:
positive (+) ions or **cations**
  These atoms have lost 1 or more electrons.

negative (−) ions or **anions**
  These atoms have gained 1 or more electrons.
  That applies to **binary** compounds.

Formation of Ionic Compounds

Ionic bonds are formed by the attraction of cations for anions usually to form solids.
Commonly, metals react with nonmetals to form ionic compounds.
The formation of NaCl is one example of an ionic compound formation.   NaCl (s)

Formation of Ionic Compounds

Reaction of Group IA Metals with Group VIIA Nonmetals

IA metal VIIA nonmetal

$2\ Li_{(s)}\ +\ F_{2(g)}$
silver   yellow
solid   gas

Formation of Ionic Compounds

Reaction of Group IA Metals with Group VIIA Nonmetals

IA metal VIIA nonmetal

$2Li_{(s)}$ + $F_{2(g)}$ → $2 LiF_{(s)}$
silver   yellow   white solid
solid   gas   with an 842°C
                    melting point

Formation of Ionic Compounds

Reaction of Group IA Metals with Group VIIA Nonmetals

The underlying reason for the formation of LiF lies in the electron configurations of Li and F.

	1s	2s	2p	
Li	↑↓	↑		loses one electron
F	↑↓	↑↓	↑↓ ↑↓ ↑	gains one electron

These atoms form ions with these configurations.

| Li+ | ↑↓ | | | same configuration as [He] |
| F− | ↑↓ | ↑↓ | ↑↓ ↑↓ ↑↓ | same configuration as [Ne] |

Formation of Ionic Compounds

Reaction of Group IA Metals with Group VIIA Nonmetals

• We can also use Lewis dot formulas to represent the neutral atoms and the ions they form.

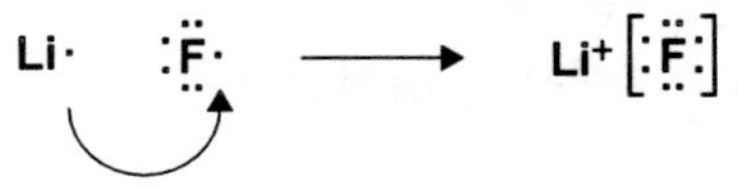

Li· :F̈· ⟶ Li+ [:F̈:]

Formation of Ionic Compounds

For the reaction of IA metals with VIA nonmetals, a good example is the reaction of lithium with oxygen.

The reaction equation is:

$$2\ Li_{(s)} + 1/2O_{2(g)} \rightarrow Li^{+}_{2}O^{2-}_{(s)}$$

Formation of Ionic Compounds

Draw the electronic configurations for Li, O, and their appropriate ions.

	2s	2p			2s	2p
Li [He]	↑		→Li$^+$			
O [He]	↑↓	↑↓ ↑ ↑	→O^{2-}		↑↓	↑↓ ↑↓ ↑↓

Draw the Lewis dot formula representation of this reaction.

Formation of Ionic Compounds

Draw the electronic representation of the Ca and N reaction.

	4s	4p			4s	4p
Ca [Ar]	↑↓		→Ca^{2+}			
	2s	2p			2s	2p
N [He]	↑↓	↑ ↑ ↑	→N^{3-}		↑↓	↑↓ ↑↓ ↑↓

$$3 \text{ Ca:} + 2 \cdot \ddot{\text{N}} \cdot \longrightarrow 3 \text{ Ca}^{2+} \; 2 \left[:\ddot{\text{N}}: \right]^{3-}$$

Other IIA and VA elements behave similarly. Symbolically, this reaction can be represented as:

$$3 \text{ M}_{(s)} + 2 \text{ X}_{(g)} \rightarrow \text{M}_3{}^{2+} \text{X}_2{}^{3-}$$

M can be the IIA elements Be to Ba
X can be the VA elements N to As.

Formation of Ionic Compounds

Simple Binary Ionic Compounds Table

Reacting Groups	Compound General Formula	Example
IA + VIIA	MX	NaF
IIA + VIIA	MX$_2$	BaCl$_2$
IIIA + VIIA	MX$_3$	AlF$_3$
IA + VIA	M$_2$X	Na$_2$O
IIA + VIA	MX	BaO
IIIA + VIA	M$_2$X$_3$	Al$_2$S$_3$

Ion Attraction and Lattice Energy

Ionic Bond

- electrostatic attraction between oppositely charged ions

- non-directional

- strength directly proportional to charges of ions and inversely proportional to distance between ion centers

Coulombic Ion Attraction and Lattice Energy

$$E \propto \frac{Q_c Q_a}{d}$$

$E = C \times \dfrac{Q_c \times Q_a}{d}$ is negative because of charges

E = ionic bond strength C (a constant) $= \dfrac{1}{4\pi\varepsilon_0}$

Q_c = charge on cation (is positive)

Q_a = charge on anion (is negative)

d = distance between centers of ions

$$E \propto \frac{Q_c Q_a}{d}$$

			Product of charges
NaCl	Na^+	Cl^-	-1
$CaCl_2$	Ca^{2+}	Cl^-	-2
CaS	Ca^{2+}	S^{-2}	-4
Al_2S_3	Al^{3+}	S^{2-}	-6

$Q_c \times Q_a$ is the heaviest factor.
Product is much more negative for Al_2S_3.

The least negative product is NaCl's

$$E \propto \frac{Q_c Q_a}{d}$$

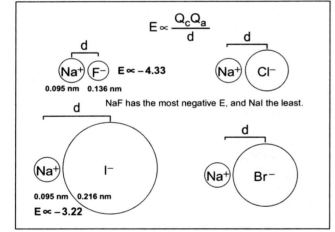

$E \propto -4.33$ (Na$^+$ F$^-$, 0.095 nm, 0.136 nm)

NaF has the most negative E, and NaI the least.

$E \propto -3.22$ (Na$^+$ I$^-$, 0.095 nm, 0.216 nm)

Lattice Energy ($\Delta H_{lattice}$)

the energy of formation of 1 mol of solid crystalline ionic compound when ions in the gas phase combine. **It is always negative.**
Among several compounds, the one with the most negative $\Delta H_{lattice}$ is said to have the highest lattice energy (absolute value)… and highest T_f

$$Mg^{2+}{}_{(g)} + 2F^-{}_{(g)} \rightarrow MgF_{2(s)} \quad \Delta H_{lattice} = -2910 \text{ KJ/mol}$$

for NaF, $\Delta H_{lattice} = -911$ KJ/mol
for KF, $\Delta H_{lattice} = -815$ KJ/mol
WHY? charge of $Mg^{2+} > Na^+ = K^+$, # of F^- (2 and 1)
$\qquad$ d, radius of $Mg^{2+} < Na^+ < K^+$ (see chapt. 8)

Lattice Energy

Small ions with high ionic charges have more negative lattice energies.
Large ions with small ionic charges have less negative lattice energies.

Use this information, plus the periodicity rules from Chapter **8**, to arrange these compounds in order of <u>decreasing values of lattice energy:</u> KCl, Al_2O_3, CaO

Ionic radii: 1.33 1.81   1.06 1.40   0.57   1.40 Å

$\Delta H_{lattice}$ $K^+Cl^- >$ $Ca^{2+}O^{2-} >$ $2Al^{3+}3O^{2-}$
　　　　　　least negative　　　　most negative

$|\Delta H_{lattice}|$ (absolute values) $Al_2O_3 > CaO > KCl$

Lattice Energy

Arrange the following compounds in order of decreasing lattice energy (least negative to most): CaO, MgO, SrO, and BaO
The anion (O^{2-}) is common to the four oxides. In the equation

$$E = C \times \frac{Q_c \, Q_a}{d}$$

Q_c and Q_a are the same (+2 and −2). Then, the difference is made by d. The four cations are in the same group, so d will be for $Ba^{2+}>Sr^{2+}>Ca^{2+}>Mg^{2+}$.
E is negative and inversely proportional to d:
the least negative　　　　　　the most negative

$Ba^{2+}O^{2-} > Sr^{2+}O^{2-} > Ca^{2+}O^{2-} >Mg^{2+}O^{2-}$ (highest E)

About Lattice Energy

According to its definition, lattice energy is negative for any compound. Please look at the examples (kJ/mol are the units):

	KBr	SrCl2	MgO
E_{lat}	− 671 >	−2127 >	−3795
$|E_{lat}|$	671 <	2127 <	3795

where $|E_{lat}|$ means absolute value (magnitude) of lattice energy.

The two rows in the table can be read as:
 MgO has the most negative lattice energy or the highest magnitude of lattice energy (highest absolute value).
 KBr has the least negative lattice energy or the lowest magnitude of lattice energy (lowest absolute value).

Ionic Bond Formation

• Born-Haber Cycle
Breaks formation of an ionic compound
from its elements into a series of theoretical
steps and considers the energetics of each.

Example: consider formation of NaCl:
$Na_{(s)} + 1/2Cl_{2(g)} \rightarrow NaCl_{(s)}$ $\Delta H° = -410.9$ KJ

The Born-Haber Cycle is a level diagram that applies Hess's law (Chapter 6):
ΔH of reaction $= \Sigma \, \Delta H$ of all theoretical steps

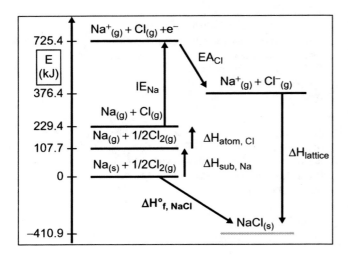

$Na^+_{(g)} + Cl_{(g)} + e^-$

E (kJ)

725.4

376.4

229.4

107.7

0

−410.9

IE_{Na}

EA_{Cl}

$Na^+_{(g)} + Cl^-_{(g)}$

$Na_{(g)} + Cl_{(g)}$

$Na_{(g)} + 1/2Cl_{2(g)}$

$Na_{(s)} + 1/2Cl_{2(g)}$

$\Delta H_{atom, Cl}$

$\Delta H_{sub, Na}$

$\Delta H_{lattice}$

$\Delta H^\circ_{f, NaCl}$

$NaCl_{(s)}$

The Born-Haber Cycle of NaCl

$\Delta H^\circ_f = \Delta H_{sublimation, Na} + \Delta H_{atom, Cl} + IE_{1, Na} + \mathbf{EA_{Cl}} + \mathbf{\Delta H_{lattice, NaCl}}$ **(the last two are negative)**

$\Delta H^\circ_f = 107.7kJ + 121.7kJ + 496.0kJ + (-349.0kJ) + (-787.3kJ)$

$\Delta H^\circ_f = -410.9kJ$

If unknown, $\Delta H_{lattice}$ can be calculated:

$\Delta H_{lattice} = \Delta H^\circ_f - \Delta H_{sublimation, Na} - \Delta H_{atom, Cl} - IE_{1,Na} - EA_{Cl}$

Practice Given the information below, determine the lattice energy of $MgCl_2$

$Mg(s) \rightarrow Mg(g)$ $\Delta H_{1\,f}^\circ = +147.1$ kJ/mol
$2\{^1/_2Cl_2(g) \rightarrow Cl(g)\}$ $2\cdot\Delta H_{2\,f}^\circ = 2(+122$ kJ/mol$)$
$Mg(g) \rightarrow Mg^+(g)$ $\Delta H_{3\,f}^\circ = +738$ kJ/mol
$Mg^+(g) \rightarrow Mg^{2+}(g)$ $\Delta H_{4\,f}^\circ = +1450$ kJ/mol
$2\{Cl(g) \rightarrow Cl^-(g)\}$ $2\cdot\Delta H_{5\,f}^\circ = 2(-349$ kJ/mol$)$
$\underline{Mg^{2+}(g) + 2\ Cl^-(g) \rightarrow MgCl_2(s)\ \Delta H^\circ_{\text{lattice energy}} = ?\ kJ/mol}$
$Mg(s) + Cl_2(g) \rightarrow MgCl_2(s)$ $\Delta H_{6\,f}^\circ = -641$ kJ/mol

The 2 in front of the $1/2Cl_2$ dissociation is due to the fact we need 2 moles of Cl atoms. The 2 also multiplies the respective ΔH (122 kJ). The same 2 is in front of the Cl Electron Affinity (−349 kJ). Contrary to Na, with only 1 e⁻ lost, Mg loses 2e⁻. Due to this, we have to write the two ionization equations for Mg, $IE_1 = 738$ kJ, $IE_2 = 1450$ kJ.

$\Delta H_{6f}^\circ = \Delta H_{1f}^\circ + 2\Delta H_{2f}^\circ + \Delta H_{3f}^\circ + \Delta H_{4f}^\circ + 2\Delta H_{5f}^\circ + \Delta H^\circ(\text{lattice energy})$

$\Delta H^\circ(\text{lattice energy}) = \Delta H_{6f}^\circ - (\Delta H_{1f}^\circ + 2\Delta H_{2f}^\circ + \Delta H_{3f}^\circ + \Delta H_{4f}^\circ + 2\Delta H_{5f}^\circ)$

$\Delta H^\circ(\text{lattice energy}) = (-641kJ) - ((+147.1kJ) + 2(+122kJ) + (+738kJ) + (+1450kJ) + 2(-349kJ))$

$\Delta H^\circ(\text{lattice energy}) = -2522kJ$

Covalent Bonding

Covalent bonds are formed when atoms **share** electrons.
If the atoms share 2 electrons (a pair), a **single covalent bond** is formed.
If the atoms share 4 electrons (two pairs), a **double covalent bond** is formed.
If the atoms share 6 electrons (three pairs), a **triple covalent bond** is formed.

The attraction between the nuclei and the electrons is electrostatic in nature. **Directional.** The atoms have a lower potential energy when bound.

Formation of Covalent Bonds

A figure shows the potential energy of an H_2 molecule as a function of the distance between the two H atoms.

Formation of Covalent Bonds

Representation of the formation of an H_2 molecule from H atoms

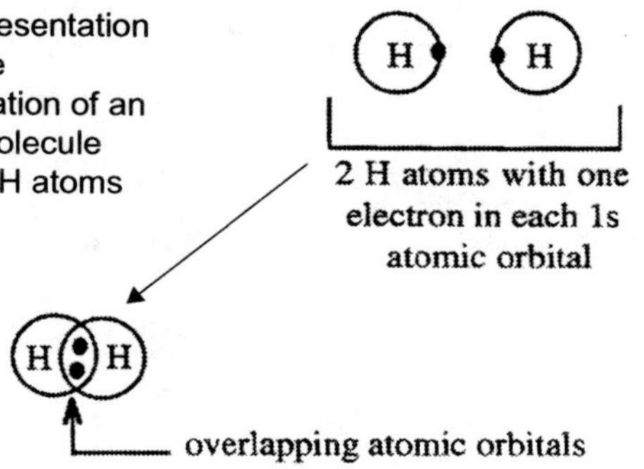

H●   ●H

2 H atoms with one electron in each 1s atomic orbital

H●H

— overlapping atomic orbitals

Formation of Covalent Bonds

We can use Lewis dot formulas to show covalent bond formation.

- H molecule formation representation.

  $$H\cdot + H\cdot \longrightarrow H:H \quad \text{or } H_2$$
  duet rule

- HCl molecule formation

  $$H\cdot + \cdot\ddot{C}l\!: \longrightarrow H\!:\!\ddot{C}l\!: \text{ or HCl}$$
  octet rule (for Cl)

Formation of Covalent Bonds

Homonuclear diatomic molecules

$$H:H \quad \text{or} \quad H-H \qquad :\ddot{F}:\ddot{F}: \quad \text{or} \quad :\ddot{F}-\ddot{F}:$$

$$:N:::N: \quad \text{or} \quad :N\equiv N:$$

Heteronuclear diatomic molecules: hydrogen halides

$$H:\ddot{F}: \quad \text{or} \quad H-\ddot{F}: \qquad H:\ddot{C}l: \quad \text{or} \quad H-\ddot{C}l:$$

$$H:\ddot{B}r: \quad \text{or} \quad H-\ddot{B}r:$$

F–F bond formation

two F atoms with one unpaired electron in each $2p$ atomic orbital

head on (maximum) overlapping of two $2p$ atomic orbitals

Writing Lewis Formulas: The Octet Rule

The **octet rule** states that representative elements usually attain stable noble gas electron configurations in *most* of their compounds.

Lewis dot formulas are based on the octet rule.

We need to distinguish between **bonding** (or **shared**) electrons and **nonbonding** (or unshared or **lone pairs of**) electrons.

The Octet Rule

N-A = S rule　　**N A S**

　Simple mathematical relationship to help us write Lewis dot formulas.

N = number of electrons **needed** to achieve a noble gas configuration.
　N usually has a value of 8 for representative elements.
　N has a value of 2 for H atoms.

A = number of electrons **available** in valence shells of the atoms.
　A is equal to the periodic group number for each element.
　A is equal to 8 for the noble gases.

S = number of electrons **shared** in bonds (in **bonding pairs.**)

A-S = number of electrons in unshared, **lone,** pairs.

The Octet Rule

For ions, we must adjust the number of electrons available, A.

　Add one e^- to A for each negative charge.

　Subtract one e^- from A for each positive charge.

The central atom in a molecule or polyatomic ion is determined by:

The atom that requires the largest number of e^-s to complete its octet. It goes in the center.

H is never central atom; it shares two e^-s only.

For two atoms in the same periodic group, the less electronegative element goes in the center (the one toward the bottom of the group.)

Lewis Structures of Covalent Compounds and Polyatomic Ions

Drawing Lewis structures by this method, use the following as a guide:

a) Draw skeletal Lewis structure.

b) Draw the Lewis electron dot structure for each atom. (Use the method in which the electrons are spread to all four sides of an imaginary square before being paired.) For the sake of keeping the drawing as neat as possible, direct single electrons on adjacent atoms toward each other.

c) Draw a line from a single unpaired electron on the central atom to a single unpaired electron on the surrounding atom. (This constitutes the formation of a covalent bond) Continue doing this until each atom has an octet (exceptions are H, Be, B, Al, elements on rows 3, 4, 5, and 6). No electrons should be left unpaired (only in rare cases will a species contain an unpaired electron). For those atoms that can have more than an octet, if all of its single electrons are used in a covalent bond, and there are surrounding atoms with electrons still to be paired, then lone electron pairs are used in bonding. The electron pair(s) being shared must be placed between the two atoms forming the bond.

d) For polyatomic ions,
−1 charge, add 1 electron to the most electronegative atom.
−2 charge, add 2 electrons (one to each of the most electronegative atoms).
−3 charge, add 3 electrons (one to each of the most electronegative atoms).
+1 charge, remove 1 electron from the least electronegative atom.

Lewis Structures of Covalent Compounds and Polyatomic Ions

e) Coordinate Covalent or Dative Bond Formation (a bond formed when both of the electrons in a bond are supplied by the same atom): If all the single electrons of one atom (atom 1) are used in bonding and the adjacent atom (atom 2) has single electrons which need to be shared, then the electrons on atom 2 (the one still having single electrons) are paired and atom 1 donates a pair of electrons to atom 2 thus forming a coordinate covalent bond. A coordinate covalent bond is represented by →. The arrowhead is pointed to the atom to which the electron pair has been donated.

Exceptions to Octet:

a) Some atoms have less than an octet. (Be, B, and Al are metals, but they can form covalent compounds.)

$$H- \qquad -Be- \qquad \overset{|}{-B-} \qquad \overset{|}{-Al-}$$

b) Central atoms of elements of rows 3, 4, 5, and 6 of the periodic table can have more than an octet due to the availability of low lying d-orbitals. These elements can have up to 18 electrons surrounding them.

A Quick Method for Drawing the Lewis Structure of Covalent Compounds and Ions

The method will be illustrated by drawing the Lewis structure of SO_3.

1) Find the total number of valence electrons by adding the number of valence electrons from each atom in the formula.

$$1S = 1(6) = 6$$
$$3O = 3(6) = \underline{18}$$
$$24$$

(For polyatomic ions, add (anions) or subtract (cations) the charge to or from the total number of valence electrons.)

2) Draw a single bond from the central atom to each of the atoms surrounding it.

$$\overset{O}{\underset{|}{O-S-O}}$$

3) Subtract the electrons used so far in the structure from the total number of electrons. Each bond contains two electrons.

$$3 \text{ bonds} \times 2\ e^- = 6\ e^-$$
$$24 - 6 = 18\ e^-$$

4) The remaining electrons are spread, as pairs, to the surrounding atoms first. Each surrounding atom (except H) will receive enough electron pairs to have an octet. If any electrons are left over, give them to the central atom.

$$\overset{..}{\underset{..}{:O:}}$$
$$:\overset{..}{O}-\overset{..}{S}-\overset{..}{O}:$$

A Quick-N-Dirty Method for Drawing the Lewis Structure of Covalent Compounds and Ions

5) If the central atom does not have an octet, then one or two single bonds are converted to double bonds or one double bond is converted to a triple bond. If possible, make two double bonds before making one triple bond.

$$:\overset{..}{O}-\overset{\overset{..}{:O:}}{\underset{..}{S}}-\overset{..}{O}: \qquad \rightarrow \qquad :\overset{..}{O}=\overset{\overset{..}{:O:}}{S}-\overset{..}{O}:$$

Remember exceptions to the octet rule (H, Be, B, Al, and elements from rows 3, 4, 5, and 6).

Lewis Structures of Covalent Compounds and Polyatomic Ions

Example: Write Lewis dot and dash formulas for hydrogen cyanide, HCN.

A = 1 (H) + 4 (C) + 5 (N) =   10   H−C−N

S (so far) = 2 bonds × 2 =   4

electrons to be spread = 10 − 4 = 6

We put the 6e⁻ over N,   H−C−N̈:

But, to fulfill the octet of C and N,   H:C:::N:

or   H−C ≡ N: with four **BP** and one **LP** over N

Lewis Structures of Covalent Compounds and Polyatomic Ions

Example: Write Lewis dot and dash formulas for the sulfite ion, SO_3^{2-}.

A = 6(S) + 3 × 6(O) + 2(−charge) = 26

S (so far) = 3 bonds × 2 =   6

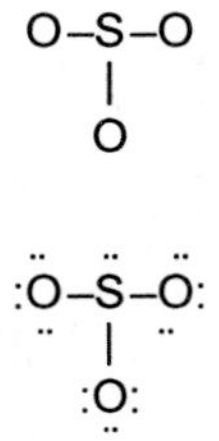

O−S−O
|
O

e⁻s to be spread = 26 − 6 = 20

     with an overall 2⁻ charge

:Ö−S−Ö:
|
:Ö:

Thus, this polyatomic ion has **3 BP** and **10 LP**.

Lewis Structures of Covalent Compounds and Polyatomic Ions

Other examples:

CO_2, NO_2^+, NO_3^-, HNO_3, SO_4^{2-}, H_2SO_4,

PO_4^{3-}, H_3PO_4

Cl_2O_7, CH_3CH_2OH (ethanol), CH_3COOH (acetic acid)

PCl_3, PCl_5, SF_6, $SOCl_2$, IF_5, IF_4^+

Lewis Structures of Covalent Compounds and Polyatomic Ions

Other examples:

BBr_3, BBr_4^-

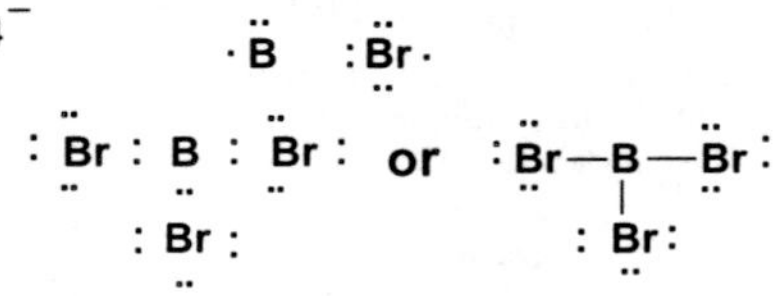

AsF_5

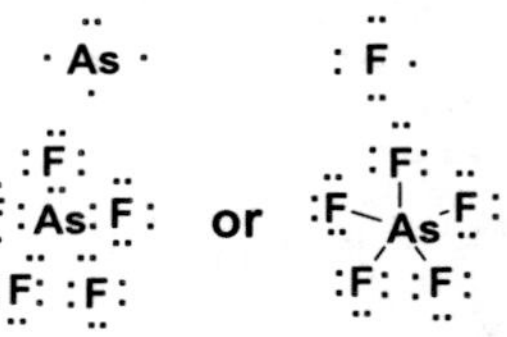

Lewis Structures of Covalent Compounds and Polyatomic Ions

Isoelectronic Species:

Molecules and/or ions having the same number of valence electrons and the same Lewis structures.

$[:N\equiv O:]^+$ $[:N\equiv N:]$ $[:C\equiv O:]$ $[:C\equiv N:]^-$

They all have **3 BP** and **2 LP**.

Resonance

Example: Write Lewis dot and dash formulas for sulfur trioxide, SO_3.

N = 8 (S) + 3 × 8 (O)	= 32
A = 6 (S) + 3 × 6 (O)	= <u>24</u>
S	= 8
A–S	= 16 **4BP, 8 LP**

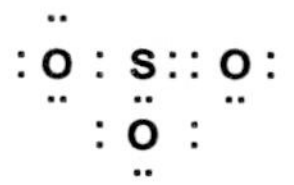

:O : S :: O :   or   :Ö—S═O :
    : O :   :Ö :

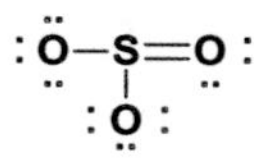

Resonance

There are three possible structures for SO_3. The double bond can be placed in one of three places.

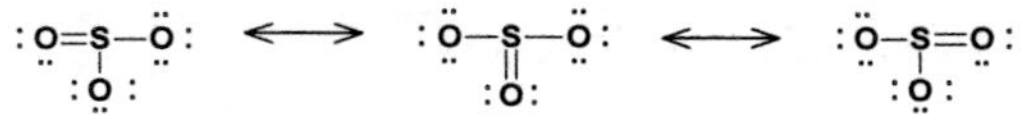

:O═S—Ö: ⟷ :Ö—S—Ö: ⟷ :Ö—S═O:
    :Ö: :Ö: :Ö:

When two or more Lewis formulas are necessary to show the bonding in a molecule, we must use equivalent resonance structures to show the molecule's structure.

Double-headed arrows are used to indicate resonance formulas.

Resonance

Resonance is a flawed method of representing molecules.

- There are no single or double bonds in SO_3.
- In fact, all of the bonds in SO_3 are equivalent.

The best Lewis formula of SO_3 that can be drawn is:

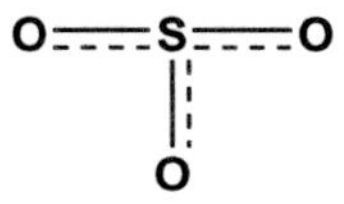

O----S----O
      |
      O

CO_2

$$:\!\ddot{O}-C-\ddot{O}\!:$$

C	4 e$^-$
2 O	12 e$^-$
	16 e$^-$
2 bonds	$-$ 4 e$^-$
	12 e$^-$
	$-$ 12 e$^-$
	0 e$^-$

Three Resonance structures for CO_2

$$:\ddot{O}=C=\ddot{O}:$$

$$:\!O\equiv C=\ddot{O}\!:$$

$$:\!\ddot{O}-C\equiv O\!:$$

Resonance

Other examples:

O_3, C_6H_6 (benzene), NO_3^- (nitrate ion)

Molecules with odd number of electrons

NO has 11 valence electrons

$$\cdot\ddot{N}=\ddot{O}$$

NO_2 has 17 valence electrons

$$:\!\ddot{O}-\dot{N}=\ddot{O} \qquad \leftrightarrow \qquad \ddot{O}=\dot{N}-\ddot{O}\!:$$

They are members of a family of substances called **Free Radicals**: they have an unpaired e$^-$. Very reactive: e.g., dimerization of NO_2 to N_2O_4.

$$2\ NO_2(g)\ \rightarrow\ N_2O_4(g)$$

Charge Distribution in Covalent Bonds and Molecules

The way the electrons are distributed in the molecule is called its **charge distribution.**

Formal Charges on Atoms:

an accounting tool for electron ownership
= (# valence e^- in free atom) – (# e^- in lone pairs on atom) – ½(# bonded e^- on atom)

Formal charge = FC =
group number of atom – [LPE + ½(BE)]

FC = group number – LPE – BP

Formal Charge (FC) and Best Structure

Best structure has:

- zero FC on all atoms
- lowest FC possible
- negative FC on most electronegative atoms, and positive FC on least electronegative atoms

The **most electronegative** elements are at the **top right of periodic table** (except for the noble gases.)

_____

_____

_____

_____

For example, consider thiocyanate ion SCN⁻

$$\left[\overset{..}{\underset{..}{N}} = C = \overset{..}{\underset{..}{S}} \right]^- \quad \left[\overset{..}{\underset{..}{C}} = S = \overset{..}{\underset{..}{N}} \right]^- \quad \left[\overset{..}{\underset{..}{S}} = N = \overset{..}{\underset{..}{C}} \right]^-$$

For example, consider thiocyanate ion SCN⁻ What is the best structure?

$$\left[\overset{..}{\underset{..}{N}} = C = \overset{..}{\underset{..}{S}} \right]^- \quad \left[\overset{..}{\underset{..}{C}} = S = \overset{..}{\underset{..}{N}} \right]^- \quad \left[\overset{..}{\underset{..}{S}} = N = \overset{..}{\underset{..}{C}} \right]^-$$

$FC_N = 5 - 4 - 2 = -1$

_____

_____

For example, consider thiocyanate ion SCN⁻

$$\left[\ddot{\text{N}} = \text{C} = \ddot{\text{S}}\right]^- \quad \left[\ddot{\text{C}} = \text{S} = \ddot{\text{N}}\right]^- \quad \left[\ddot{\text{S}} = \text{N} = \ddot{\text{C}}\right]^-$$

$FC_N = 5 - 4 - 2 = -1$

$FC_C = 4 - 0 - 4 = 0$

$FC_S = 6 - 4 - 2 = 0$

For example, consider thiocyanate ion SCN⁻

$$\overset{-1}{\left[\ddot{\text{N}}\right.} = \overset{0}{\text{C}} = \overset{0}{\left.\ddot{\text{S}}\right]}^- \quad \left[\ddot{\text{C}} = \text{S} = \ddot{\text{N}}\right]^- \quad \left[\ddot{\text{S}} = \text{N} = \ddot{\text{C}}\right]^-$$

$FC_N = 5 - 4 - 2 = -1$

$FC_C = 4 - 4 - 2 = -2$

$FC_S = 6 - 0 - 4 = +2$

For example, consider thiocyanate ion SCN⁻

$$\overset{-1}{\left[\ddot{\text{N}}\right.} = \overset{0}{\text{C}} = \overset{0}{\left.\ddot{\text{S}}\right]}^- \quad \overset{-2}{\left[\ddot{\text{C}}\right.} = \overset{+2}{\text{S}} = \overset{-1}{\left.\ddot{\text{N}}\right]}^- \quad \left[\ddot{\text{S}} = \text{N} = \ddot{\text{C}}\right]^-$$

$FC_N = 5 - 0 - 4 = +1$

$FC_C = 4 - 4 - 2 = -2$

$FC_S = 6 - 4 - 2 = 0$

For example, consider thiocyanate ion SCN⁻

$$\overset{-1}{\left[\ddot{\text{N}}\right.} = \overset{0}{\text{C}} = \overset{0}{\left.\ddot{\text{S}}\right]}^- \quad \overset{-2}{\left[\ddot{\text{C}}\right.} = \overset{+2}{\text{S}} = \overset{-1}{\left.\ddot{\text{N}}\right]}^- \quad \overset{0}{\left[\ddot{\text{S}}\right.} = \overset{+1}{\text{N}} = \overset{-2}{\left.\ddot{\text{C}}\right]}^-$$

For example, consider thiocyanate ion SCN⁻

$$\left[\overset{-1}{\underset{\cdot\cdot}{\ddot{N}}} = C = \overset{0}{\underset{\cdot\cdot}{\ddot{S}}}\right]^{-} \quad \left[\overset{-2}{\underset{\cdot\cdot}{\ddot{C}}} = \overset{+2}{S} = \overset{-1}{\underset{\cdot\cdot}{\ddot{N}}}\right]^{-} \quad \left[\overset{0}{\underset{\cdot\cdot}{\ddot{S}}} = \overset{+1}{N} = \overset{-2}{\underset{\cdot\cdot}{\ddot{C}}}\right]^{-}$$

best structure because lowest FC and −FC on most electronegative atom

Sum of formal charges = −1 + 0 + 0 = −1

Formal Charge (FC)

NH_4^+

$\begin{array}{c} H \\ | \\ H-N-H \\ | \\ H \end{array}$

$N = 5 - 0 - 4 = +1$

$H = 1 - 0 - 1 = 0$

SO_4^{2-}

CO_2 $\ddot{O}=C=\ddot{O}$

$S = 6 - 0 - 4 = -2$

$O = 6 - 6 - 1 = -1$

$C = 4 - 0 - 4 = 0$ $O = 6 - 4 - 2 = 0$

O_3 (ozone) $:\ddot{O}-O=\ddot{O}$ $\ddot{O}=O-\ddot{O}:$

Practice: Identify Structures with Better or Equal Resonance Forms and Draw Them

CO_2

all 0 $\ddot{O}=C=\ddot{O}$

H_3PO_4

P = +1
rest 0

$SeOF_2$

Se = +1

SO_3^{2-}

S = +1

NO_2^-

P_2H_4

all 0

Practice: Identify Structures with Better or Equal FC Resonance Forms and Draw Them

CO_2

none $\ddot{O}=C=\ddot{O}$

H_3PO_4 all 0

$SeOF_2$

SO_3^{2-}

S ≈ 0
in all new
res. forms

NO_2^-

P_2H_4 none

173

Central Atoms with Single-Bond Pairs and Lone Pairs

Valence Shell Electron-Pair Repulsion theory (VSEPR Theory):

Regions of high electron density around the central atom are arranged as far apart as possible to minimize repulsions.

Lone pairs of electrons require more volume than shared pairs. Hence,

- Lone pair to lone pair is the strongest repulsion.
- Lone pair to bonding pair is intermediate repulsion.
- Bonding pair to bonding pair is the weakest repulsion.

Mnemonic for repulsion strengths:

lp/lp> lb/bp > bp/bp

VSEPR Theory

There are five basic molecular shapes based on the number of regions of high electron density around the central atom.

Several modifications of these five basic shapes will also be examined.

VSEPR Theory

1. Two regions of high electron density (**BP+LP**) around the central atom. Linear, 180°

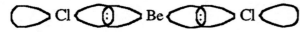

2. Three regions of high electron density (**BP+ LP**) around the central atom. Trigonal planar, 120°

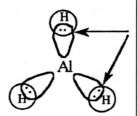

VSEPR Theory

3. Four regions of high electron density (**BP+LP**) around the central atom. Tetrahedral, 109.5°

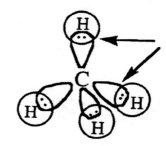

VSEPR Theory

4. Five regions of high electron density (**BP + LP**) around the central atom. Trigonal bipyramidal, 90°, 120°, 180°

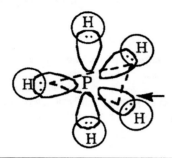

VSEPR Theory

5. Six regions of high electron density (**BP + LP**) around the central atom. Octahedral, 90°, 180°

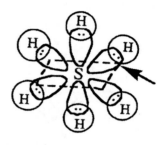

VSEPR Theory

Frequently, we will describe two geometries for each molecule.

Electronic geometry is determined by the locations of regions of high electron density around the central atom(s).

Molecular geometry is determined by the arrangement of atoms around the central atom(s).

> ***Electron pairs are not used in the molecular geometry determination; just the positions of the atoms in the molecule are used.***

VSEPR Theory

An example of a molecule that has the same electronic and molecular geometries is methane - CH_4.

Electronic and molecular geometries are tetrahedral.

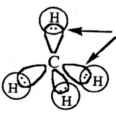

VSEPR Theory

An example of a molecule that has different electronic and molecular geometries is ammonia - NH_3.

Electronic geometry is tetrahedral.

Molecular geometry is trigonal pyramidal.

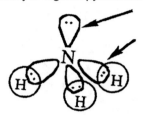

Predicting Molecular Geometries and Bond Angles

Write the Lewis structure.

Count the **number of bonding pairs and lone pairs around the central atom (regions of high electron density).**

Treat a double or triple bond as one **region h.e.d.**

Use Table provided to predict geometry. When lone pairs are present, bond angles are slightly less than those given (except, linear = 180° and squared planar = 90°). When only two atoms are present, it is inappropriate to speak of bond angles.

Predicting Molecular Geometries and Bond Angles

2 BP	CO_2, CS_2, $BeCl_2$	linear
3 BP	BF_3, CO_3^{2-}	trigonal planar
2 BP 1 LP	SO_2, NOCl	bent
4 BP	CH_4, NH_4^+, BF_4^-	tetrahedral
3 BP 1 LP	NH_3, H_3O^+	trigonal pyramidal
2 BP 2 LP	H_2O, NH_2^-	V-shaped or bent
5 BP	PCl_5, AsF_5	trigonal bipyramid
6 BP	SF_6, PCl_6^-	octahedral

Bond Polarity and Electronegativity
Polar and Nonpolar Covalent Bonds

Covalent bonds in which the electrons are shared equally are designated as **nonpolar** covalent bonds.

Nonpolar or **pure covalent** bonds have a symmetrical charge distribution.

To be pure covalent, the two atoms involved in the bond must be the same element to share equally.

H : H or H-H : N : : : N : or : N≡N :

Pure Covalent Bonds

2 H atoms with one electron in each 1s atomic orbital

overlapping atomic orbitals

Bond Polarity and Electronegativity
Polar and Nonpolar Covalent Bonds

Covalent bonds in which the electrons are **not shared equally** are designated as **polar** covalent bonds.

Polar covalent bonds have an asymmetrical charge distribution.

To be a polar covalent bond, the two atoms involved in the bond must have different electronegativities.

Electronegativity

Electronegativity is a measure of the relative tendency of an atom to attract electrons to itself when *chemically combined with another element*.

Electronegativity is measured on the **Pauling scale.**

Fluorine is the most electronegative element.

Cesium and francium are the least electronegative elements.

For the representative elements, electronegativities usually **increase** from left to right across periods and **decrease** from top to bottom within groups.

Electronegativity

Example: Arrange these elements based on their electronegativity.

Se, Ge, Br, As   (period # 4)

Ge < As < Se < Br

Arrange these elements based on their electronegativity.

Be, Mg, Ca, Ba   (group IIA)

Ba < Ca < Mg < Be

Electronegativity, polarity, and Ionic bonds

A **non-polar** bond is formed when the difference of electronegativity is between **0.0 and 0.4.**

A **polar** bond is formed when the difference of electronegativity is **0.5 – 1.8 (partial charge).**

Ionic bonds form when the difference of electronegativity is **greater than 1.8 (net charge).**

Cl is more electronegative

no charge	δ^+ δ^-	
Se – H (0.3)	C – Cl (0.5)	K$^+$ and Br$^-$ (1.9)
non-polar	**polar**	**ionic**

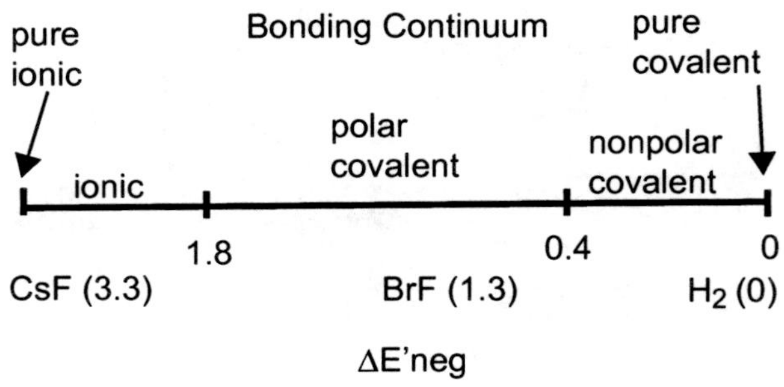

Bonding Continuum

pure ionic → ... pure covalent

ionic | polar covalent | nonpolar covalent

1.8   0.4   0

CsF (3.3)   BrF (1.3)   H$_2$ (0)

ΔE'neg

Molecular Polarity

The higher the difference of electronegativity, the more polar the bond is:

H   I

Electroneg ativities 2.1   2.5

0.4

Difference = 0.4 slightly polar bond

H   F

Electroneg ativities 2.1   4.0

1.9

Difference = 1.9 very polar bond

Polar and Nonpolar Covalent Bonds

Polar molecules can be attracted by magnetic and electric fields.

Dipole Moments

Molecules whose centers of positive and negative charge do not coincide have an asymmetric charge distribution and are polar. These molecules have a **dipole moment**.

The dipole moment has **the symbol m**; μ is the product of the distance, d, separating charges of equal magnitude and opposite sign, and the magnitude of the charge, δ.

It is indicated by crossed arrow pointing from positive end to negative end of dipole

$$\mathbf{m} = \delta \times d \qquad \longmapsto$$

Dipole Moments

Molecules that have a small separation of charge have a small μ.

Molecules that have a large separation of charge have a large μ.

For example, HF and HI:

	δ^+ H - Fδ^-	δ^+ H - Iδ^-	Ratio
	1.91 Debye units	**0.38 Debye units**	**5.0**
ΔE'neg	2.0	0.1	20

Dipole Moments

There are some <u>nonpolar</u> molecules that have <u>polar</u> bonds.

There are two conditions that must be true for a molecule to be polar.

1. There must be at least one polar bond present or one lone pair of electrons.

2. The polar bonds, if there are more than one, and lone pairs must be arranged so that their dipole moments do *not* cancel one another.

Polarity of Molecules

Molecules in which dipole moments of the bonds do not cancel are polar molecules

$$\mu_{total} \neq 0$$

Molecules that do not contain polar bondsor in which all dipole moments cancel are non-polar molecules

$$\mu_{total} = 0$$

CO$_2$ vs H$_2$O

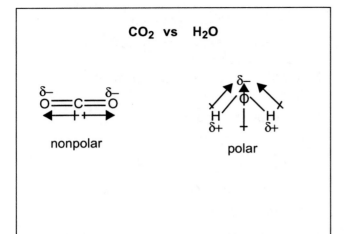

nonpolar

polar

Dipole Moments of Polyatomic Molecules

Some models

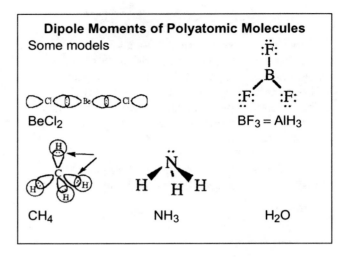

BeCl$_2$

BF$_3$ = AlH$_3$

CH$_4$ NH$_3$ H$_2$O

Dipole Moments of Polyatomic Molecules

Some models

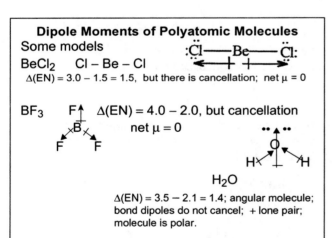

BeCl$_2$ Cl – Be – Cl

Δ(EN) = 3.0 – 1.5 = 1.5, but there is cancellation; net μ = 0

BF$_3$ F Δ(EN) = 4.0 – 2.0, but cancellation

net μ = 0

H$_2$O

Δ(EN) = 3.5 – 2.1 = 1.4; angular molecule;
bond dipoles do not cancel; + lone pair;
molecule is polar.

Dipole Moments of Polyatomic Molecules

Some examples to work...

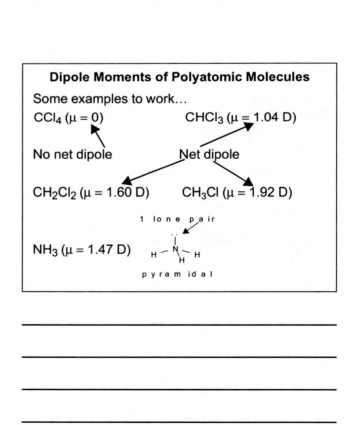

CCl$_4$ (μ = 0)   CHCl$_3$ (μ = 1.04 D)

No net dipole   Net dipole

CH$_2$Cl$_2$ (μ = 1.60 D)   CH$_3$Cl (μ = 1.92 D)

1 lone pair

NH$_3$ (μ = 1.47 D)   H – N – H
                                              H

pyramidal

CH₄ ... CCl₄   Polar or Not?

CH₄	CH₃Cl	CH₂Cl₂	CHCl₃	CCl₄
μ = 0	μ = 1.92 D	μ = 1.60 D	μ = 1.04 D	μ = 0
No net dipole	Net dipole	Net dipole	Net dipole	No net dipole

There is no cancellation.

Only CH₄ and CCl₄ are NOT polar. These are the only two molecules that are "symmetrical." The individual dipoles cancel out.

Bonding Order

Order of a bond is the number of bonding electron pairs shared by two atoms in a molecule. There are bond orders of 1 (single), 2 (double), 3 (triple), and fractional (1.5, 2.5, ...).

$$\text{Avg. bond order} = \frac{\text{\# of shared pairs linking atoms}}{\text{\# of links in the molecule or ion (no L.P.)}}$$

H – Cl	O – O = O	O = C = O	H – C ≡ N
1	1 2	2 2	1 3

For the molecule: (**around the central atom**):
B.O.   1/1 = 1   3/2 = 1.5   4/2 = 2   4/2 = 2

Bond Strength and Bond Length
bond length
(shortest) triple < double < single

bond strength
single < double < triple (strongest)

Bond	Length(Å)	Strength (kJ/mol)
C – C	1.54	348
C = C	1.34	614 (**not 2x**)
C ≡ C	1.20	839 (**not 3x**)

Bond Energies

- Chemical reactions involve breaking bonds in reactant molecules and making new bonds to create the products.
- The $\Delta H°_{reaction}$ can be calculated by comparing the cost of breaking old bonds to the profit from making new bonds.
- The amount of energy it takes to break one mole of a bond in a compound is called the **bond energy**.
  - in the gas state
  - homolytically – each atom gets ½ of bonding electrons

Trends in Bond Energies

• The more electrons two atoms share, the stronger the covalent bond.
  – C≡C (837 kJ) > C=C (611 kJ) > C–C (347 kJ)
  – C≡C (891 kJ) > C=N (615 kJ) > C–N (305 kJ)
• The shorter the covalent bond, the stronger the bond.
  – Br–F (237 kJ) > Br–Cl (218 kJ) > Br–Br (193 kJ)
  – bonds get weaker down the column of periodic table.

Using Bond Energies to Estimate $\Delta H°_{rxn}$

• The actual bond energy depends on the surrounding atoms and other factors.
• We often use **average bond energies** to estimate the ΔH_{rxn}
  —works best when all reactants and products in gas state.
• Bond breaking is endothermic, ΔH(breaking) = +
• Bond making is exothermic, ΔH(making) = –

$$\Delta H_{rxn} = \Sigma\ (\Delta H(\text{bonds broken})) + \Sigma\ (\Delta H(\text{bonds}$$
$$\text{all} > 0 \qquad\qquad \text{all} < 0\ \text{formed}))$$

Bond Lengths

Table of bond energy

Estimate the Enthalpy of the Following Reaction

H — H + O = O ⟶ H — O — O — H

$H_2(g) + O_2(g) \longrightarrow H_2O_2(g)$

Reaction involves breaking 1mol H-H and 1 mol O=O and making 2 mol H-O and 1 mol O-O.

bonds broken (energy cost)
(+436 kJ) + (+498 kJ) = +934 kJ
bonds made (energy released)
2(–464 kJ) + (–142 kJ) = –1070
$\Delta H_{rxn} = (+934\ \text{kJ}) + (–1070.\ \text{kJ}) = –136\ \text{kJ}$

(Appendix $\Delta H°_f = –136.3$ kj/mol)

Calculating the DH_{combustion} of acetic acid(g)

Calculating the $\Delta H_{combustion}$ of acetic acid(g)

$$CH_3COOH(g) + 2O_2(g) \rightarrow 2H_2O(g) + 2CO_2(g)$$

$$
\begin{array}{c}
\quad\;\; H \quad O\!-\!H \\
\quad\;\; | \quad\;\; | \\
H\!-\!C\!-\!C\!=\!O \quad + \quad 2\,O\!=\!O \rightarrow 2H\!-\!O\!-\!H \; + \; 2\,O\!=\!C\!=\!O \\
\quad\;\; | \\
\quad\;\; C
\end{array}
$$

$\Delta H_{comb} = 3E_{C-H} + E_{C-C} + 2E_{O=O} + E_{C=O} + E_{C-O} + E_{O-H}$
$$\qquad\qquad - 4E_{C=O}(\text{* in CO}_2) - 4E_{O-H}$$

$\Delta H_{comb} = 3 \times 414 + 347 + 2 \times 498 + 736 + 360 - \mathbf{4 \times 799}$
$$\qquad\quad - 3 \times 464$$

Bond Lengths

- The distance between the nuclei of bonded atoms is called the **bond length.**
- Because the actual bond length dependsontheotheratomsaround the bond, we often use the **average bond length.**
  - averaged for similar bonds from many compounds

Trends in Bond Lengths

- The more electrons two atoms share, the shorter the covalent bond:
  - $C \equiv C$ (120 pm) $<$ $C=C$ (134 pm) $<$ C–C (154 pm)
  - $C \equiv N$ (116 pm) $<$ $C=N$ (128 pm) $<$ C–N (147 pm)
- Decreases from left to right across period:
  - C–C (154 pm) $>$ C – N (147 pm) $>$ C–O (143 pm)
- Increases down the column:
  - F–F (144 pm) $<$ Cl – Cl (198 pm) $<$ Br–Br (228 pm)
- In general, **as bonds get longer, they also get weaker.**

Chapter 10

Bonding and Molecular Structure: Orbital Hybridization and Molecular Orbitals

Goals

- Understand the differences between valence bond theory and molecular orbital theory.

- Identify the hybridization of an atom in a molecule or ion.

- Understand the differences between bonding and antibonding molecular orbitals.

- Write the molecular orbital configuration for simple diatomic molecules.

Orbitals and Bonding Theories

VSEPR Theory only explains molecular shapes. It says nothing about bonding in molecules.

In **Valence Bond (VB) Theory (Linus Pauling)**: atoms share electron pairs by allowing their atomic orbitals to overlap.

Another approach to rationalize chemical bonding is the **Molecular Orbital (MO) Theory (Robert Mulliken)**: molecular orbitals are spread out or "delocalized" over the molecule.

Valence Bond (VB) Theory

Covalent bonds are formed by the **overlap** of atomic orbitals.

Atomic orbitals on the central atom can mix and exchange their character with other atoms in a molecule.

Process is called **hybridization.**

Hybrids are common:

Pink flowers
Mules

Hybrid orbitals have the same shapes as predicted by VSEPR.

185

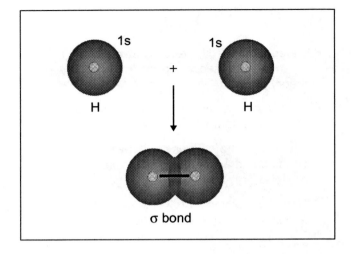

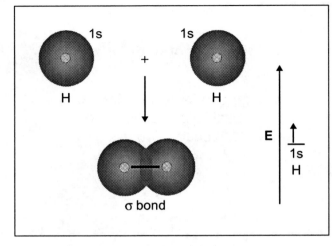

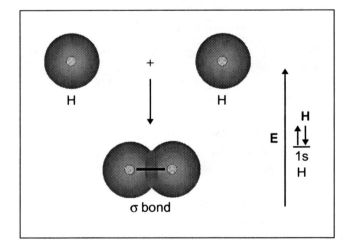

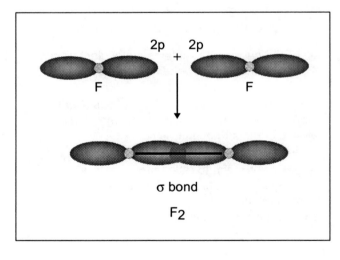

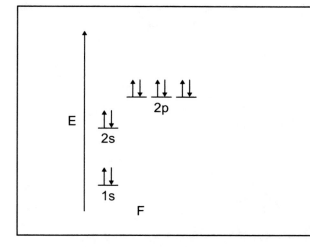

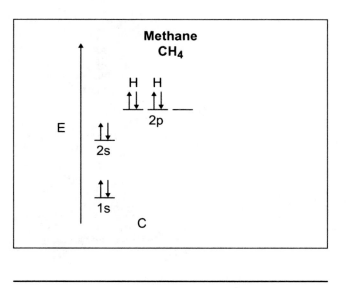

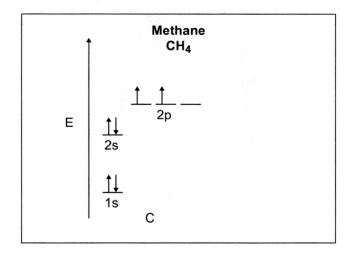

Methane
CH$_4$

E

H H
↑↓ ↑↓ ___
2p

H$^+$
↑↓
2s

↑↓
1s

C

Methane
CH$_4$

E

H H H$^-$
↑↓ ↑↓ ↑↓
2p

H$^+$
↑↓
2s

↑↓
1s

C

Methane
CH$_4$

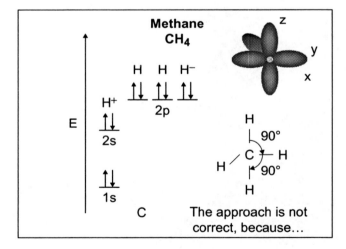

E

H H H$^-$
↑↓ ↑↓ ↑↓
2p

H$^+$
↑↓
2s

↑↓
1s

C

The approach is not
correct, because…

Methane
CH$_4$

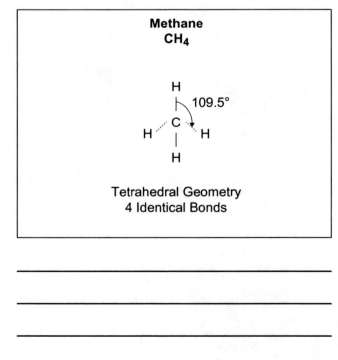

Tetrahedral Geometry
4 Identical Bonds

Problem and Solution

C must have 4 identical orbitals in valence shell for bonding.

Solution: hybridization (**theoretical mixing of the four atomic orbitals of carbon atom, the 2s and the three 2p**)

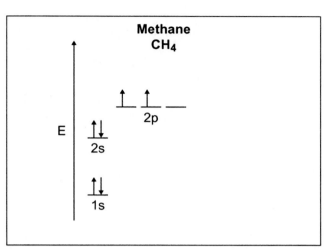

_____

_____

_____

_____

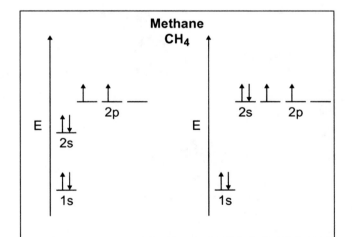

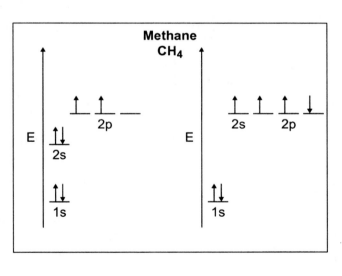

_____

_____

_____

_____

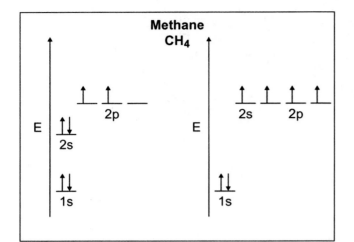

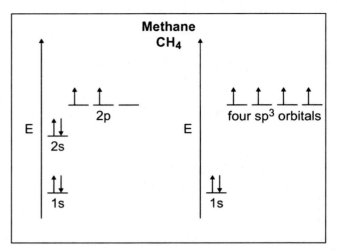

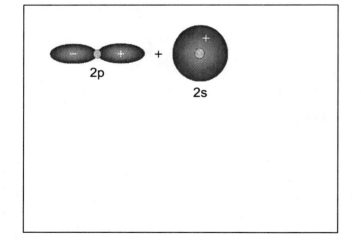

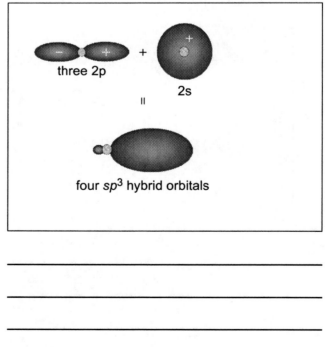

4 identical sp^3 hybrid orbitals: there are four because there was the combination of one s and three p atomic orbitals (25% s, 75% p).

tetrahedral geometry

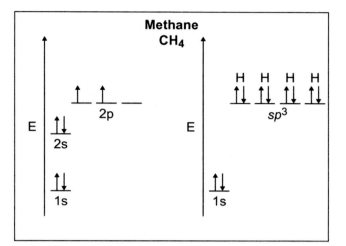

Methane
CH₄

Valence Bond (VB) Theory		
Regions of High Electron Density (BP + LP)	Electronic Geometry	Hybridization, Angles(°)
2	Linear	sp, **180**
3	Trigonal planar	sp^2 **120**
4	Tetrahedral	sp^3 **109.5**
5	Trigonal bipyramidal	sp^3d **120, 90, 180**
6	Octahedral	sp^3d^2 **90, 180**

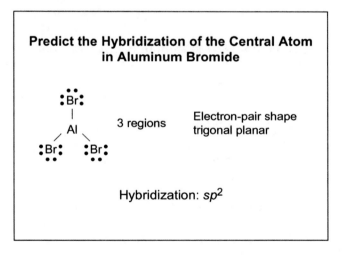

Predict the Hybridization of the Central Atom in Aluminum Bromide

3 regions

Electron-pair shape trigonal planar

Hybridization: sp^2

Trigonal Planar Electronic Geometry, sp^2

Electronic Structures: BF_3

	1s	2s	2p
B	↑↓	↑↓	↑

	1s	2s	2p	2p		1s	sp^2 hybrid		
B	↑↓	↑↓	↑	↑	⇒	↑↓	↑	↑	↑

	2s	2p
F [He]	↑↓	↑↓ ↑↓ ↑

Trigonal Planar Electronic Geometry, sp²

AlH_3, BF_3, BH_3,

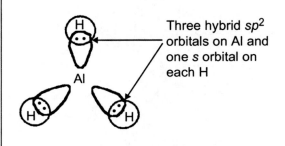

Three hybrid sp^2 orbitals on Al and one *s* orbital on each H

Predict the Hybridization of the Central Atom in Carbon Dioxide
CO_2

$$\ddot{O} = C = \ddot{O}$$

2 regions

Electron-pair shape, linear

Hybridization: *sp* (50% s, 50% p)

Linear Electronic Geometry, *sp*

Electronic Structures: $BeCl_2$

	1s	2s	2p		1s	*sp* hybrid	
Be	↑↓	↑↓		⇒	↑↓	↑	↑

	3s	3p
Cl [Ne]	↑↓	↑↓ ↑↓ ↑

Predict the Hybridization of the Central Atom in Beryllium Chloride

Two regions: electron-pair shape

sp hybridization

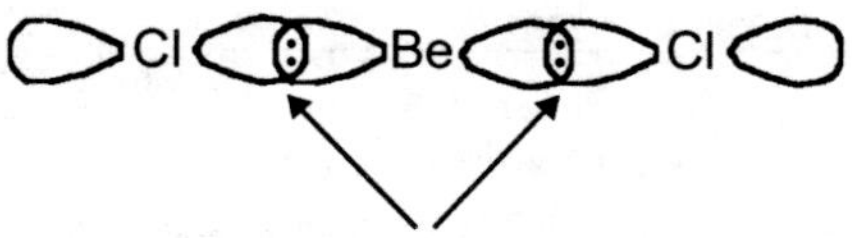

two *sp* hybrid orbitals on Be
and one *p* orbital on each Cl

Predict the Hybridization of the Central Atom in PF$_5$

Five regions: Trigonal Bipyramidal Electronic Geometry

sp^{3}d hybridization, five *sp^{3}d* hybrid orbitals

Regions of high electron density	Electronic geometry	Hybridization at central atom (angle)	Hybridized orbital orientation
5	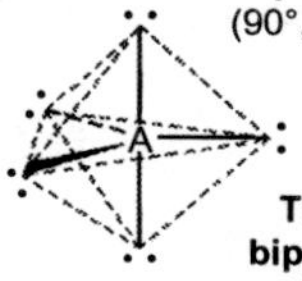 Trigonal bipyramidal	*sp^{3}d* or *dsp^3* (90°, 120°, 180°)	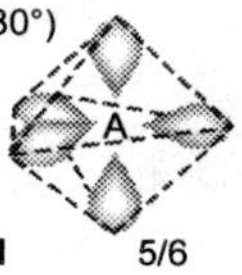 5/6

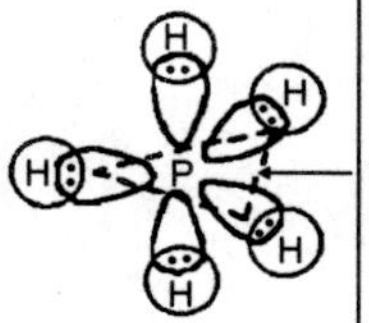

sp^{3}d hybrid orbitals on P and *s* orbital on H

Predict the Hybridization of the Central Atom in Xenon Tetrafluoride

---

Predict the Hybridization of the Central Atom in Xenon Tetrafluoride

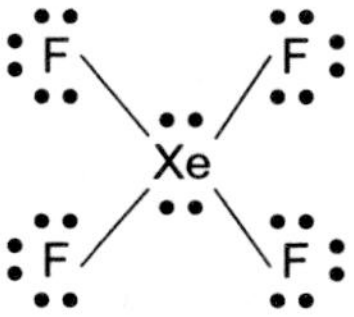

6 regions

electron-pair shape

octahedral

Predict the Hybridization of the Central Atom in Xenon Tetrafluoride

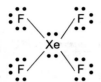

6 regions

electron-pair shape

octahedral

sp^3d^2 hybridization

Predict the Hybridization of the Central Atom in SH_6 or SF_6

Six regions: Octahedral Electronic Geometry

— sp^3d^2 hybridization,
six sp^3d^2 hybrid orbitals

Regions of high electron density	Electronic geometry	Hybridization at central atom (angle)	Hybridized orbital orientation
6		sp^3d^2 or d^2sp^3 (90°, 180°)	

Octahedral

sp^3d^2 hybrid orbitals on S and s orbital on H

6/6

Consider Ethylene, C_2H_4

Consider Ethylene, C_2H_4

H—C=C—H
with H atoms

Consider Ethylene, C₂H₄

$$H \diagdown \atop H \diagup C = C \diagup \atop \diagdown {H \atop H}$$

3 regions
trigonal planar

Consider Ethylene, C₂H₄

$$H \diagdown \atop H \diagup C = C \diagup \atop \diagdown {H \atop H}$$

3 regions
trigonal planar
sp² hybridization

Consider Ethylene, C₂H₄

$$H \diagdown \atop H \diagup C = C \diagup \atop \diagdown {H \atop H}$$

3 regions
trigonal planar
sp²
hybridization

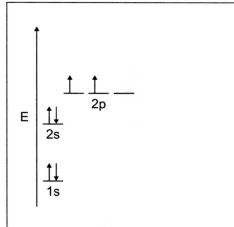

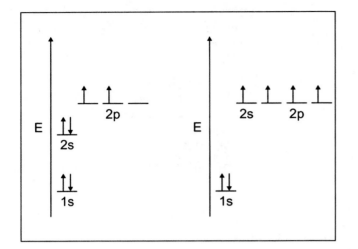

_____

_____

_____

_____

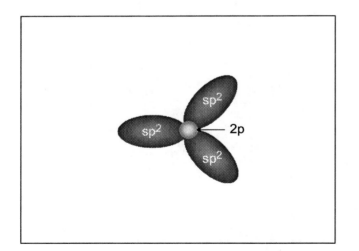

_____

_____

_____

_____

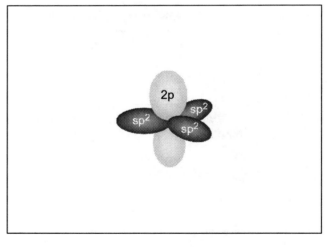

_____

_____

_____

_____

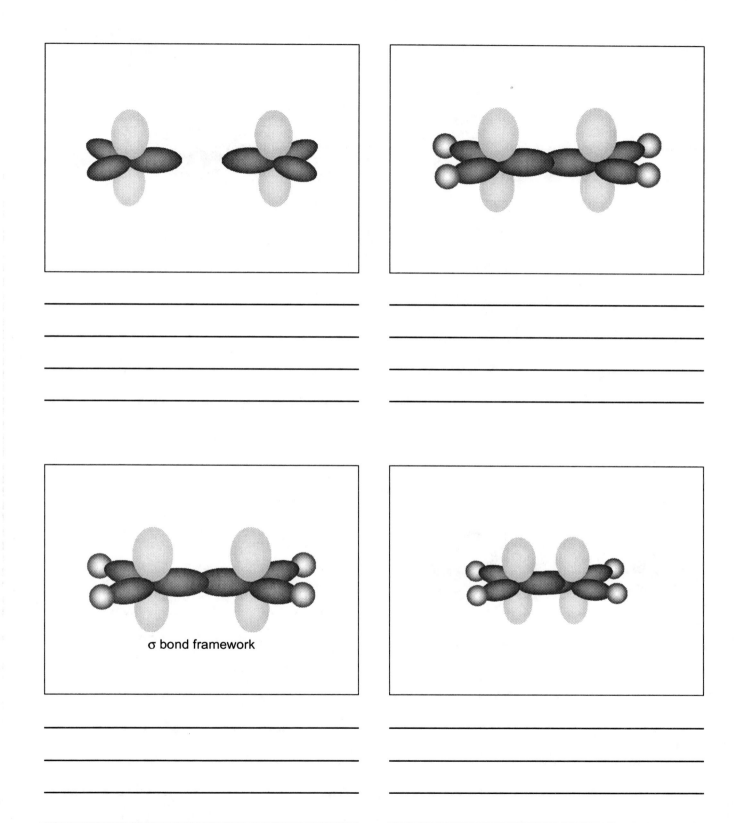

σ bond framework

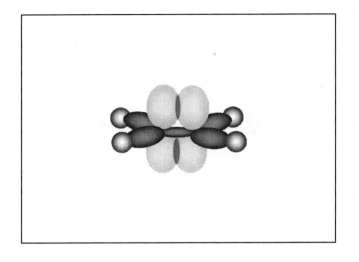

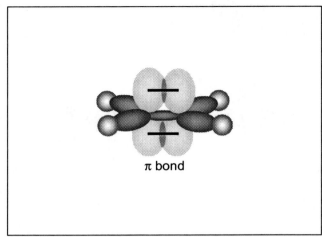

π bond

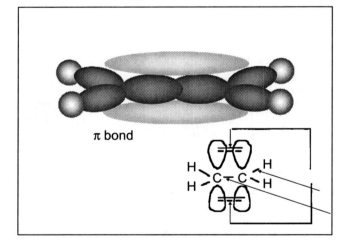

π bond

Compounds Containing Double Bonds

Thus, a C = C bond looks like this and is made of two parts, one σ and one π bond.

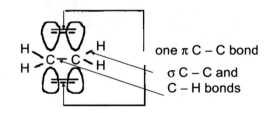

one π C – C bond

σ C – C and
C – H bonds

198

Consider Acetylene, C_2H_2

$$H - C \equiv C - H$$

Consider Acetylene, C_2H_2

$$\boxed{H - C \equiv} C - H$$

2 regions
linear

Consider Acetylene, C_2H_2

$$\boxed{H - C \equiv} C - H$$

2 regions
linear
sp hybridization

Consider Acetylene, C_2H_2

$$H - C \equiv \boxed{C - H}$$

2 regions
linear
sp hybridization

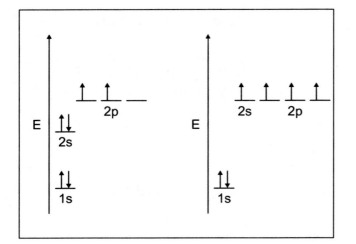

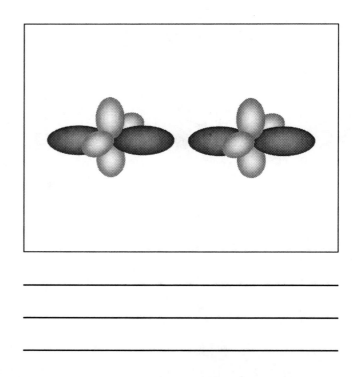

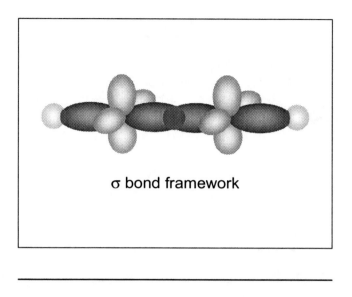

σ bond framework

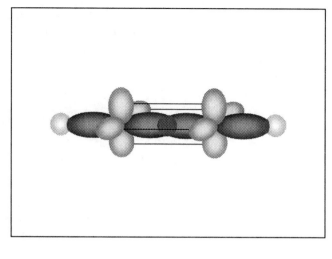

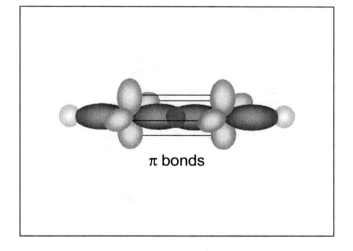

π bonds

Compounds Containing Triple Bonds

A **σ bond** results from the **head-on** overlap of two *sp* hybrid orbitals.

The unhybridized *p* orbitals form two **π bonds** (**side-on** overlap of atomic orbitals.)

Note that a triple bond consists of one **σ** and two **π bonds**.

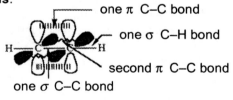

one π C–C bond
one σ C–H bond
second π C–C bond
one σ C–C bond

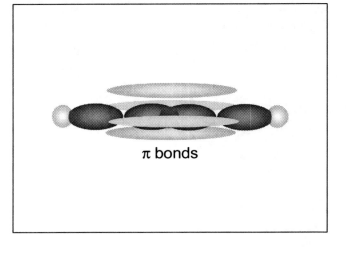

π bonds

Generally

- single bond is a σ bond
- double bond consists of 1 σ and 1 π bond
- triple bond consists of σ and 2 π bonds

Molecular Orbital (MO) Theory

- When atoms combine to form molecules, atomic orbitals overlap and are then combined to form molecular orbitals.
- # of orbitals are conserved.
- A molecular orbital is an orbital associated with more than 1 nucleus.
- Like any other orbital, an MO can hold 2 electrons.
- Consider hydrogen atoms bonding to form H_2.

Molecular Orbital Theory

- Combination of atomic orbitals on different atoms forms molecular orbitals (MOs) so that electrons in MOs belong to the molecule as a whole.

- Waves that describe atomic orbitals have both positive and negative phases or amplitudes.

- As MOs are formed the phases can interact **constructively** or **destructively**.

Molecular Orbitals

- There are two simple types of molecular orbitals that can be produced by the overlap of atomic orbitals.

  Head-on overlap of atomic orbitals produces σ **(sigma)** orbitals.

  Side-on overlap of atomic orbitals produces π **(pi)** orbitals.

- **Two 1s atomic** orbitals that overlap produce two molecular orbitals designated as:

  σ_{1s} or bonding molecular orbital

  σ^*_{1s} or antibonding molecular orbital

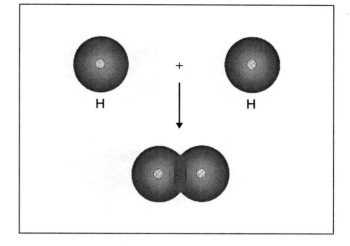

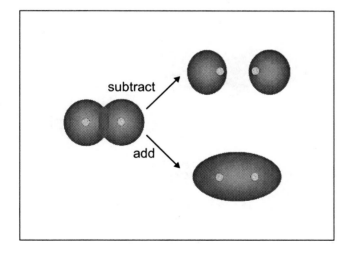

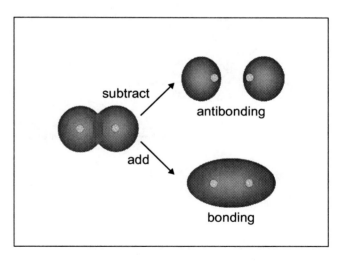

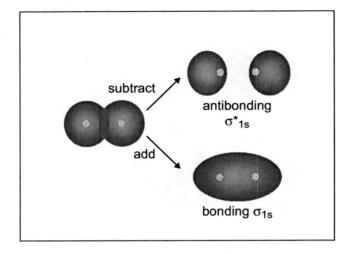

Molecular Orbital Energy Level Diagram

Now that we have seen what these MOs look like and a little of their energetics, how are the orbitals filled with electrons?
Order of filling of MOs obeys same rules as for atomic orbitals. Including:

Aufbau principle: increasing energy
Pauli's Exclusion: two unaligned e- per orbital, with opposite spins (+1/2 and −1/2)
Hund's Rule: maximum spin; unpaired electrons in degenerate orbitals have same spin (+1/2 or −1/2)

Thus, the following energy level diagram results for the homonuclear diatomic molecules H_2 and He_2.

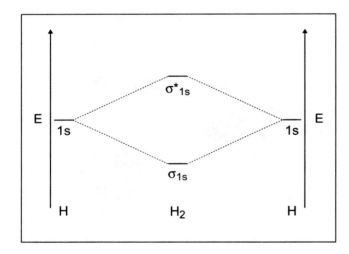

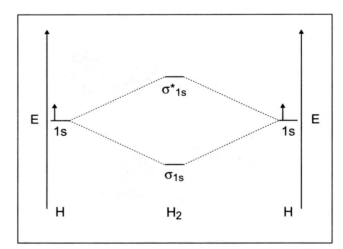

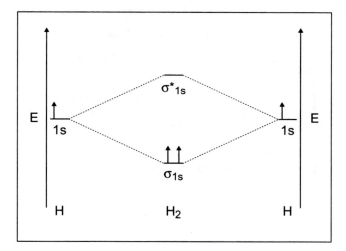

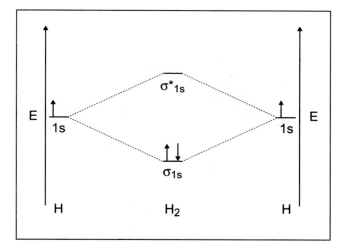

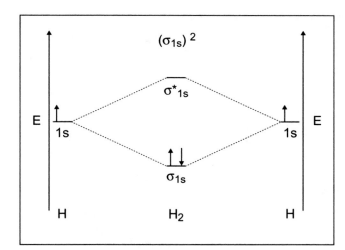

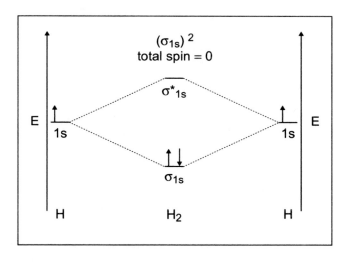

- Diamagnetic: slightly repelled by a magnetic field
  – total spin = 0
- Paramagnetic: attracted to a magnetic field
  – total spin not 0

- Bond Order = $\dfrac{\text{(bonding } e^- - \text{antibonding } e^-)}{2}$

Bond Order and Bond Stability

The larger the bond order, the more stable the molecule or ion is.

Bond order = **0** implies there are equal numbers of electrons in bonding and antibonding orbitals, ~ same stability as separate atoms: **no bond formed**

Bond order > **0** implies there are more electrons in bonding than antibonding orbitals. Molecule is more stable than separate atoms.

The greater the bond order, the shorter the bond length and the greater the bond energy.

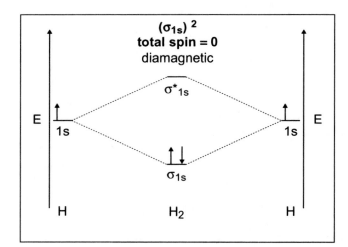

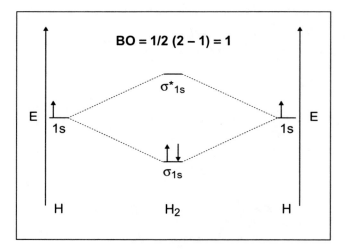

Consider He₂

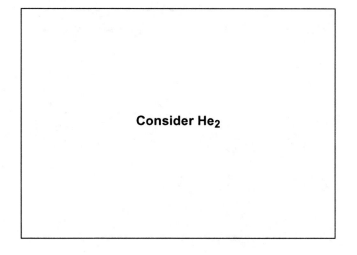

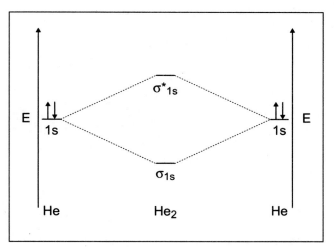

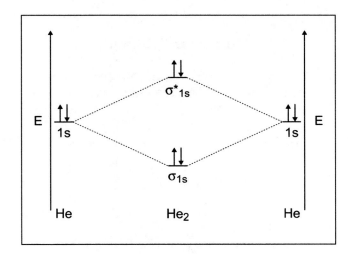

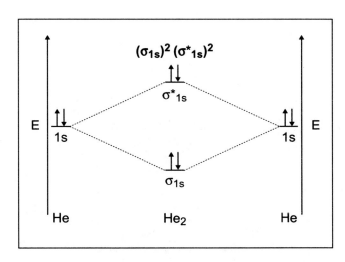

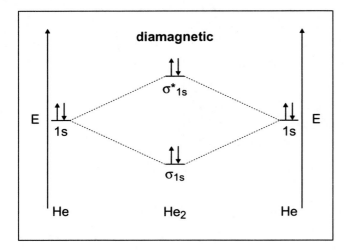

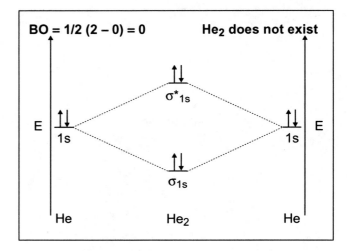

Combination of p Atomic Orbitals

Molecular Orbitals

The head-on overlap of two corresponding *p* atomic orbitalson different atoms, say $2p_x$ with $2p_x$ produces:

σ_{2px} bonding orbital

σ^*_{2px} antibonding orbital

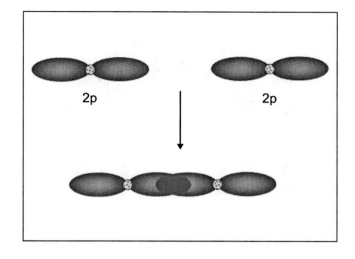

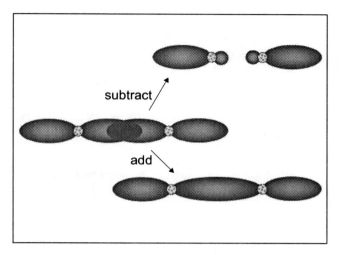

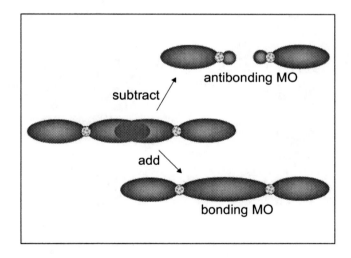

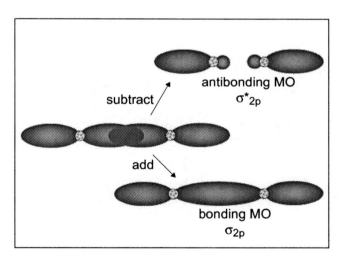

Molecular Orbitals

Side-on overlap of two corresponding p atomic orbitals on different atoms (say $2p_y$ with $2p_y$ or $2p_z$ with $2p_z$) produces:

π_{2p_y} or π_{2p_z} (both are bonding orbitals)

$\pi^*_{2p_y}$ or $\pi^*_{2p_z}$ (both are nonbonding orbitals)

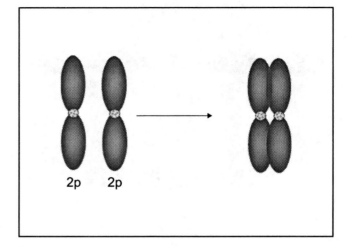

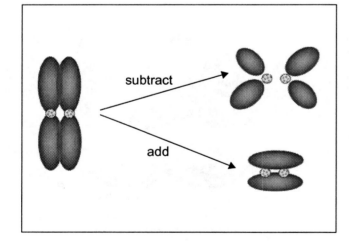

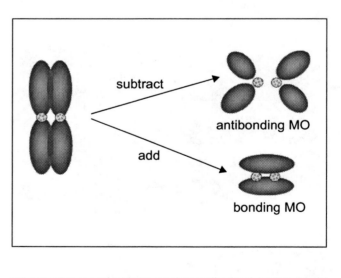

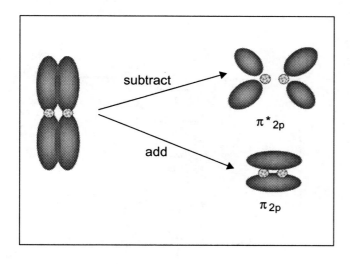

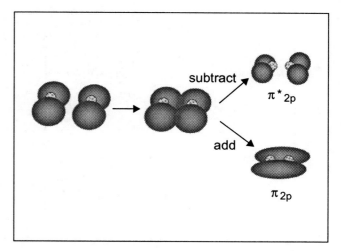

Consider Li₂

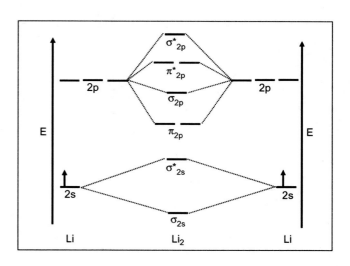

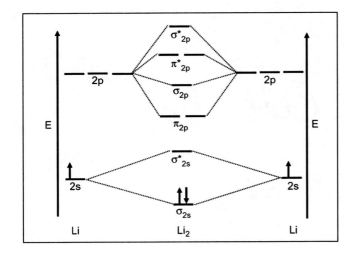

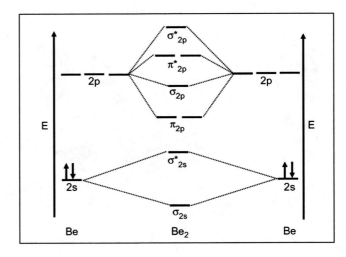

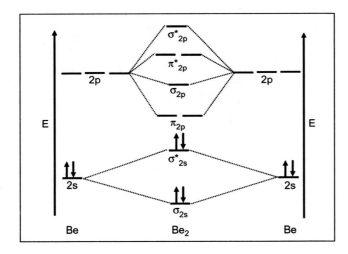

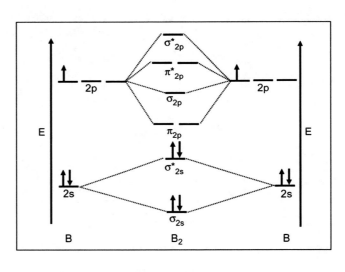

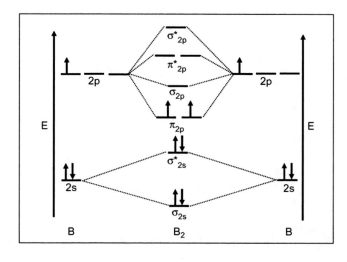

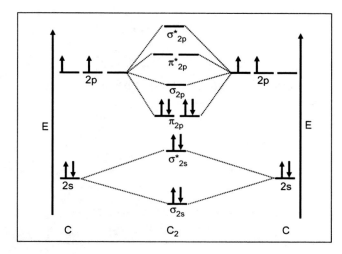

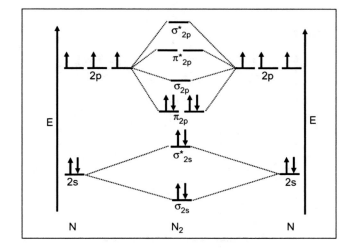

Homonuclear Diatomic Molecules

In shorthand notation, we represent the configuration of N_2 as:

$$N_2 \; \sigma_{1s}^2 \; \sigma_{1s}^{*2} \; \sigma_{2s}^2 \; \sigma_{2s}^{*2} \; \pi_{2p_y}^2 \; \pi_{2p_z}^2 \; \sigma_{2p}^2$$

Bond Order of N$_2$

$$N_2 \; \sigma_{1s}^2 \; \sigma_{1s}^{*2} \; \sigma_{2s}^2 \; \sigma_{2s}^{*2} \; \pi_{2p_y}^2 \; \pi_{2p_z}^2 \; \sigma_{2p}^2$$

The greater the bond order of a bond, the more stable we predict it to be.

For N$_2$, the bond order is:

$$bo = \frac{10 - 4}{2}$$

$$= \frac{6}{2}$$

$$= \underline{3} \text{ corresponding to a triple bond in VB theory}$$

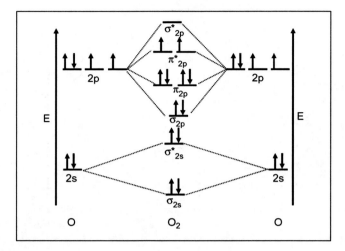

_____

_____

_____

_____

Homonuclear Diatomic Molecules

In shorthand notation, we represent the configuration of O$_2$ as:

$$O_2 \; \sigma_{1s}^2 \; \sigma_{1s}^{*2} \; \sigma_{2s}^2 \; \sigma_{2s}^{*2} \; \pi_{2p_y}^2 \; \pi_{2p_z}^2 \; \pi_{2p_x}^2 \; \pi_{2p_y}^{*1} \; \pi_{2p_z}^{*1}$$

$$bo = \frac{10 - 6}{2} = 2$$

We can see that O$_2$ is a paramagnetic molecule (two unpaired electrons).

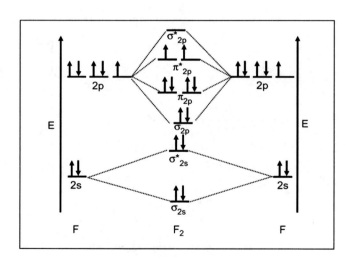

_____

_____

_____

_____

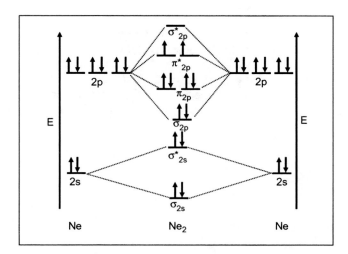

Bond Order for Ne₂

$$BO = \frac{2 \times 4 - 2 \times 4}{2} = 0$$

We can see that **Ne₂** is not stable. It does not exist.

Delocalization and Shapes of Molecular Orbitals

Molecular orbital theory describes shapes in terms of *delocalization of electrons*.
Carbonate ion (CO_3^{2-}) is a good example.
VB Theory   MO Theory

Delocalization and Shapes of Molecular Orbitals

Benzene, C_6H_6, Resonance structure - VB theory

Lewis formulas

Delocalization and Shapes of Molecular Orbitals

This is the picture of the valence bond (VB) theory.

Delocalization and Shapes of Molecular Orbitals

The structure of benzene is described well by molecular orbital theory. The three pairs of π electrons are equally shared by the six carbon atoms.

Heteronuclear Diatomic Molecules

- The more electronegative atom has lower energy orbitals.
- When the combining atomic orbitals are identical and equal energy, the weight of each atomic orbital in the molecular orbital is equal.
- When the combining atomic orbitals are different kinds and energies the atomic orbital closest in energy to the molecular orbital contributes more to the molecular orbital:
  - Lower energy atomic orbitals contribute more to the bonding MO
  - Higher energy atomic orbitals contribute more to the antibonding MO
- Nonbonding MOs remain localized on the atom donating its atomic orbitals.

Polyatomic Molecules

- When many atoms are combined together, the atomic orbitals of all the atoms are combined to make a set of molecular orbitals that are delocalized over the entire molecule.

- Gives results that better match real molecule properties than either Lewisor Valence Bond theories.

Chapter 11

An Introduction to Organic Chemistry

Types of Organic Compounds

- Vast majority of over 20 million known compounds are based on C: **organic compounds**.
- Generally contain C and H + other elements
- Great variety of compounds

$$H-\underset{Br}{\overset{H}{\underset{|}{\overset{|}{C}}}}-\overset{H}{\underset{|}{C}}=\overset{H}{\underset{|}{C}}-O-\overset{O}{\overset{\|}{C}}-\underset{\underset{H}{|}}{\overset{H}{\overset{|}{N}}}-\overset{O}{\overset{\|}{C}}-S-\underset{Cl}{\overset{H}{\underset{|}{\overset{|}{C}}}}-\overset{H}{\underset{|}{C}}\equiv H$$

Isomerism

- **Isomers** have identical composition but different structures.
- **Two forms of isomerism**
  - Constitutional (or structural)
  - Stereoisomerism
- **Constitutional**
  - Same empirical formula but different atom-to-atom connections
- **Stereoisomerism**
  - Same atom-to-atom connections but different arrangement in space.

Structural Isomers

Ethanol C_2H_6O CH_3CH_2OH $H-\overset{H}{\underset{H}{\overset{|}{\underset{|}{C}}}}-\overset{H}{\underset{H}{\overset{|}{\underset{|}{C}}}}-O-H$

Dimethyl ether C_2H_6O CH_3OCH_3 $H-\overset{H}{\underset{H}{\overset{|}{\underset{|}{C}}}}-O-\overset{H}{\underset{H}{\overset{|}{\underset{|}{C}}}}-H$

Stereoisomers: Geometric

Geometric isomers can occur when there is a $C=C$ double bond.

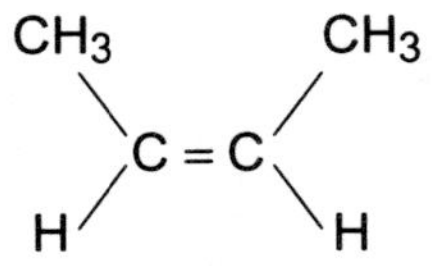

$$CH_3 \quad CH_3$$
$$\diagdown \quad \diagup$$
$$C = C$$
$$\diagup \quad \diagdown$$
$$H \qquad H$$

cis-2-butene

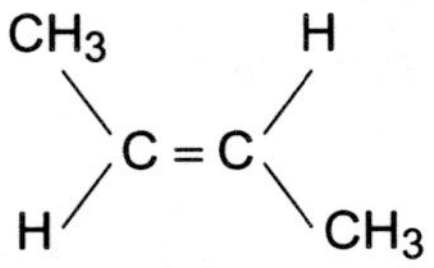

$$CH_3 \qquad H$$
$$\diagdown \quad \diagup$$
$$C = C$$
$$\diagup \quad \diagdown$$
$$H \qquad CH_3$$

trans-2-butene

Stereoisomers: Optical

- **Optical isomers** are molecules with non-superimposable mirror images.
- Such molecules are called **CHIRAL.**
- Pairs of chiral molecules are **enantiomers.**
- Chiral molecules in solution can rotate the plane of plane-polarized light.

Chiral Compounds—Polarized Light

Stereoisomers Isomers

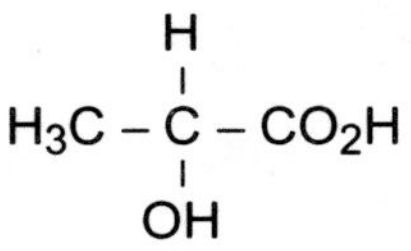

$$\begin{array}{c} H \\ | \\ H_3C - C - CO_2H \\ | \\ OH \end{array}$$

Lactic acid

Chirality generally occurs when a C atom has 4 different groups attached.

Hydrocarbons

- Alkanes have the general formula C_nH_{2n+2}
- CH_4 = methane
- C_2H_6 = ethane
- C_3H_8 = propane
- C_4H_{10} = butane

$$CH_3 - CH_2 - CH_2 - CH_3$$

- C_5H_{12} = pentane

Table 11.2 • Selected Hydrocarbons of the Alkane Family, C_nH_{2n+2}*

Name	Molecular Formula	State at Room Temperature
methane	CH_4	
ethane	C_2H_6	gas
propane	C_3H_8	
butane	C_4H_{10}	
pentane	C_5H_{12} (pent- = 5)	
hexane	C_6H_{14} (hex- = 6)	
heptane	C_7H_{16} (hept- = 7)	liquid
octane	C_8H_{18} (oct- = 8)	
nonane	C_9H_{20} (non- = 9)	
decane	$C_{10}H_{22}$ (dec- = 10)	
octadecane	$C_{18}H_{38}$ (octadec- = 18)	solid
eicosane	$C_{20}H_{42}$ (eicos- = 20)	

* This table lists only selected alkanes. Liquid compounds with 11–16 carbon atoms are also known, and there are many solid alkanes with more than 18 C atoms.

Hydrocarbons: Alkanes

Alkanes are colorless gases, liquids, and solids.

Generally unreactive (but undergo combustion).

Not polar (or low polarity) and so are not soluble in water.

Alkanes
Alkanes with Substituents

There are three substituents on the chiral C atom of lactic acid.

$$H_3C - \overset{\overset{\displaystyle H}{|}}{\underset{\underset{\displaystyle OH}{|}}{C}} - CO_2H$$

Isomers of Butane

Isomers
- have the same molecular formula
- have different atom arrangements
- of butane (C_4H_{10}) are a straight chain and a branched chain

$$H_3C - CH_2 - CH_2 - CH_3$$

$$\begin{array}{c} CH_3 \\ | \\ H_3C - CH - CH_3 \end{array}$$

Alkyl groups

Alkyl groups are:
- alkanes that are missing one H.
- substituents attached to carbon chains.
- named with a **–yl** ending.

$$\begin{array}{c} H \\ | \\ H - C - \\ | \\ H \end{array} \qquad H_3C - \qquad \textbf{methyl}$$

$$\begin{array}{cc} H & H \\ | & | \\ H - C - C - \\ | & | \\ H & H \end{array} \qquad CH_3 - CH_2 - \qquad \textbf{ethyl}$$

Naming Substituents

Names and Formulas of some substituents

In the IUPAC system,

- a carbon branch is named as an **alkyl group**.
- halogen atoms are named as **halo**.

Substituent	Name
$H_3C -$	**methyl**
$H_3C - CH_2 -$	**ethyl**
$H_3C - CH_2 - CH_2 -$	**propyl**
$H_3C - CH - CH_3$	**isopropyl**
F–, Cl–, Br–, I–	**fluoro, chloro, bromo, iodo**

Alkanes with Substituents

$$\begin{array}{c} CH_3 \\ | \\ CH_3 - CH - CH_3 \end{array} \qquad \text{methylpropane}$$

methyl groups

$$\begin{array}{cc} CH_3 & CH_3 \\ | & | \\ CH_3 - CH - CH_2 - CH - CH_3 \end{array} \quad \text{2,4-dimethylpentane}$$

220

Naming Alkanes

Give the name of

$$CH_3 - CH - CH - CH_3$$
$$\quad\quad\ |\quad\ |$$
$$\quad\quad CH_3\ CH_3$$

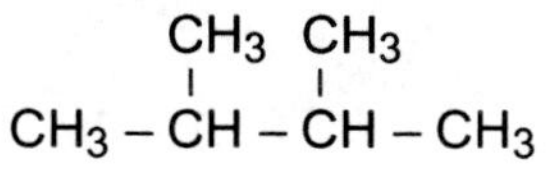

STEP 1 Name the longest continuous chain.

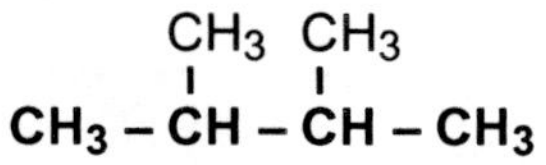

$$CH_3 - CH - CH - CH_3$$
$$\quad\quad\ |\quad\ |$$
$$\quad\quad CH_3\ CH_3$$
butane

Naming Alkanes

Give the name of

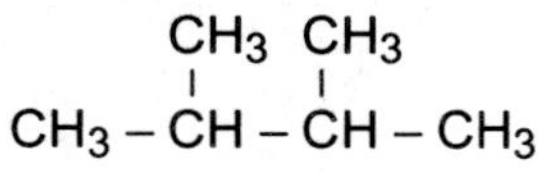

$$CH_3 - CH - CH - CH_3$$
$$\quad\quad\ |\quad\ |$$
$$\quad\quad CH_3\ CH_3$$

STEP 2 Number chain.

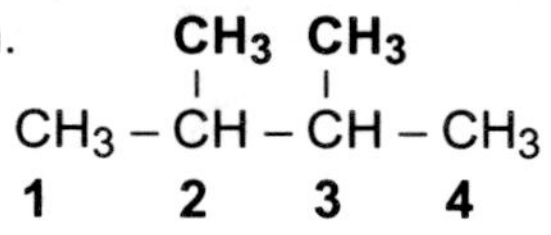

$$CH_3 - CH - CH - CH_3$$
$$\quad\quad\ |\quad\ |$$
$$\quad\quad \mathbf{CH_3\ CH_3}$$
$$\mathbf{1\quad\ 2\quad\ 3\quad\ 4}$$

STEP 3 Locate substituents and name.
2,3-dimethylbutane

_____

Problem

Write the name of:

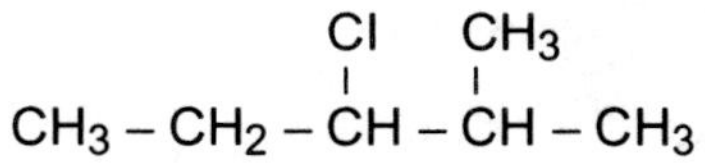

$$CH_3 - CH_2 - CH - CH - CH_3$$
$$\quad\quad\quad\quad\quad\ |\quad\ |$$
$$\quad\quad\quad\quad\ Cl\quad CH_3$$

Solution

STEP 1 Longest chain is **pentane.**

STEP 2 Number chain from end nearest substituent.

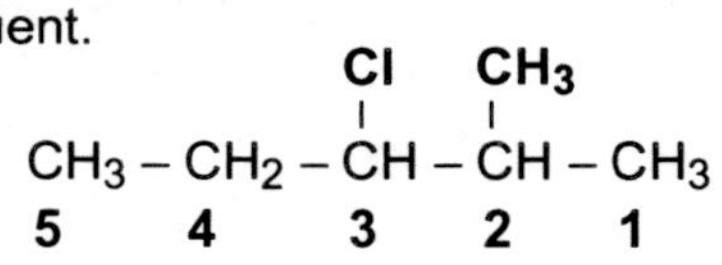

$$CH_3 - CH_2 - CH - CH - CH_3$$
$$\quad\quad\quad\quad\quad\ |\quad\ |$$
$$\quad\quad\quad\quad\ Cl\quad CH_3$$
$$\mathbf{5\quad\ 4\quad\ 3\quad\ 2\quad\ 1}$$

STEP 3 Locate substituents and name alphabetically.

3-chloro-2-methylpentane

_____

Problem

Give the IUPAC name for each of the following:

A.

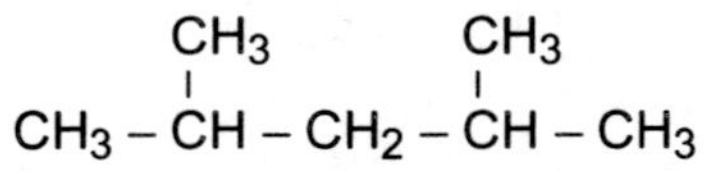

$$CH_3 - CH - CH_2 - CH - CH_3$$

with CH_3 groups on the 2nd and 4th carbons

B.

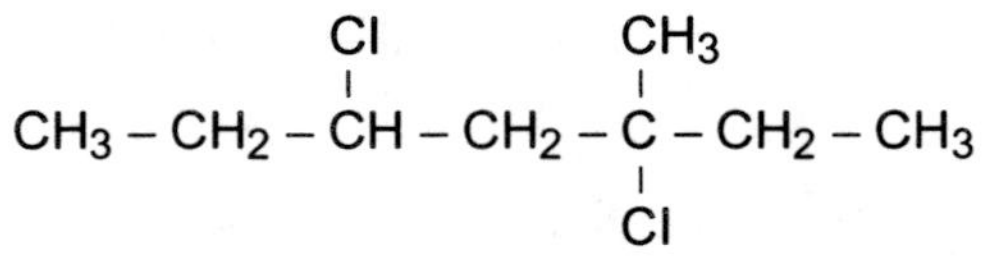

$$CH_3 - CH_2 - CH - CH_2 - C - CH_2 - CH_3$$

with Cl above the 3rd carbon (CH), CH_3 above and Cl below the central carbon

Solution

A.

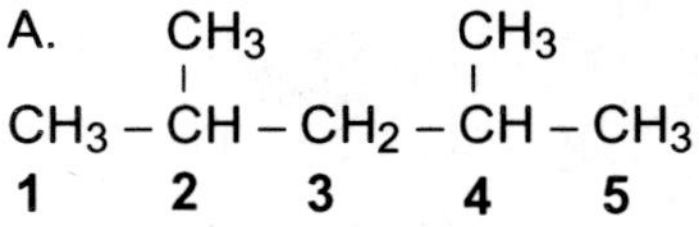

$$\underset{1}{CH_3} - \underset{2}{CH} - \underset{3}{CH_2} - \underset{4}{CH} - \underset{5}{CH_3}$$ **2,4-dimethylpentane**

with CH_3 on carbon 2 and carbon 4

B.

more substituents = C3

$$\underset{7}{CH_3} - \underset{6}{CH_2} - \underset{5}{CH} - \underset{4}{CH_2} - \underset{3}{C} - \underset{2}{CH_2} - \underset{1}{CH_3}$$

with Cl on carbon 5, CH_3 and Cl on carbon 3

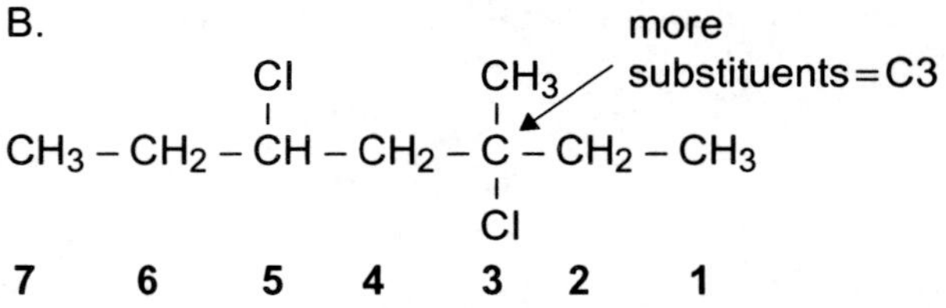

3,5-dichloro-3-methylheptane

Name the Following

A.

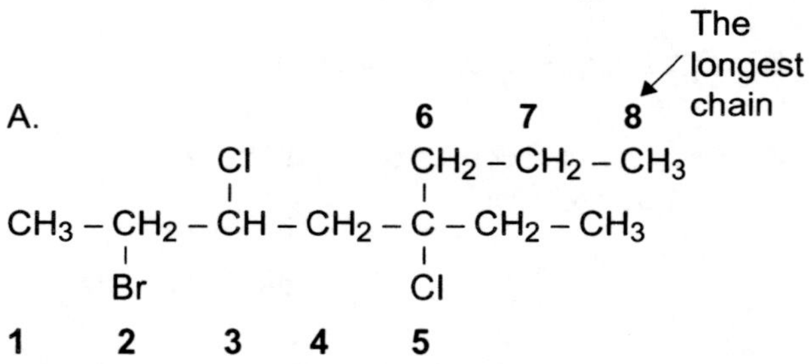

The longest chain

$$\underset{1}{CH_3} - \underset{2}{CH_2} - \underset{3}{CH} - \underset{4}{CH_2} - \underset{5}{C} - \underset{6}{CH_2} - \underset{7}{CH_2} - \underset{8}{CH_3}$$

with Cl on carbon 3, Br on carbon 2, $CH_2 - CH_2 - CH_3$ (carbons 6,7,8) and Cl on carbon 5

2-bromo-3,5-dichloro-5-ethyloctane

Practice Problem

Draw the condensed structural formula for 2-bromo-4-chlorobutane.

Solution

2-bromo-4-chlorobutane

STEP 1 Longest chain has 4 carbon atoms.

C – C – C – C

STEP 2 Number chain and add substituents.

Alphabetically:
bromo- first than chloro-

Br
|
C – C – C – C – Cl
1   2   3   4

STEP 3 Add hydrogen to complete 4 bonds
to each C.

Br
|
CH₃–CH–CH₂–CH₂–Cl

WRONG name! It is 3-bromo-1-chlorobutane

Naming Cycloalkanes with Substituents

The name of a substituent is placed in front of
The cycloalkane name.

methylcyclobutane

CH₃

chlorocyclopentane

Cl

Two Problems

Name each of the following.

1. ▷— CH₃

2. CH₂ – CH₃

Solution

Name each of the following.

1. **methyl**cyclopropane

2. **ethyl**cyclohexane

Unsaturated Hydrocarbons

Alkenes and Alkynes

Saturated Hydrocarbons

- have the maximum number of hydrogen atoms attached to each carbon atom

- are alkanes and cycloalkanes with single C-C bonds

$$CH_3 - CH_2 - CH_3$$

The cycle is considered as one unsaturation.

Unsaturated Hydrocarbons

- have fewer hydrogen atoms attached to the carbon chain than alkanes

- are alkenes with double bonds

- are alkenes with triple bonds

Ethene:

$$\begin{array}{cc} H & H \\ \diagdown & \diagup \\ C = C \\ \diagup & \diagdown \\ H & H \end{array}$$

Ethyne:

$$H - C \equiv C - H$$

Bond Angles in Alkenes and Alkynes

According to VSEPR theory:

- The three groups bonded to carbon atoms in a double bond are at 120° angles.

- Alkenes are flat because the atoms in a double bond all lie in the same plane.

- The two groups bonded to each carbon in a triple bond are at 180° angles.

Ethene:

Bond angle = 120°

Ethyne:

Bond angle = 180°

$$H - C \equiv C - H$$

Naming Alkenes

The names of alkenes:
- use the corresponding alkane name
- change the ending to **–ene**

Alkene	IUPAC	Common
$H_2C = CH_2$	eth**ene**	ethyl**ene**
$H_2C = CH - CH_3$	prop**ene**	**propylene**
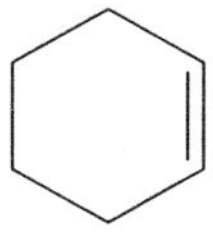	cyclohex**ene**	

Ethene (Ethylene)

Ethene or ethylene:
- is an alkene with the formula C_2H_4
- has two carbon atoms connected by a double bond
- has two H atoms bonded to each C atom
- is flat with all the C and H atoms in the same plane
- is used to accelerate the ripening of fruits

Ethene:

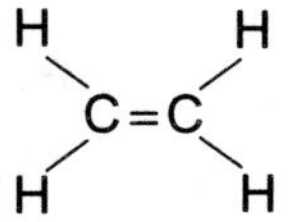

Naming Alkynes

The names of alkynes:
- use the corresponding alkane name
- change the ending to **–yne**

Alkyne	IUPAC	Common
$HC \equiv CH$	eth**yne**	acetylene
$HC \equiv C - CH_3$	prop**yne**	

Naming Alkenes and Alkynes

When the carbon chain of an alkene or alkyne has four or more C atoms, number the chain to give the lowest number to the first carbon in the double or triple bond.

$CH_2 = CH - CH_2 - CH_3$ 1-butene
 1　　 2　　 3　　 4

$CH_3 - CH = CH - CH_3$ 2-butene
 1　　 2　　 3　　 4

$CH_3 - CH_2 - C \equiv C - CH_3$ 2-pentyne
 5　　 4　　 3　 2　 1

Problem

Write the IUPAC name for each of the following:

1. $CH_2=CH-CH_2-CH_3$
2. $CH_3-CH=CH-CH_3$
3. $CH_3-CH=\underset{\underset{CH_3}{|}}{C}-CH_3$

4. $CH_3-C\equiv C-CH_3$

Solution

Write the IUPAC name for each of the following:

1. $CH_2=CH-CH_2-CH_3$ **1-butene**
2. $CH_3-CH=CH-CH_3$ **2-butene**
3. $CH_3-CH=\underset{\underset{CH_3}{|}}{C}-CH_3$ **2-methyl-2-butene**

4. $CH_3-C\equiv C-CH_3$ **2-butyne**

Problem

Write the IUPAC name for each of the following:

A. $CH_3-CH_2-C\equiv C-CH_3$

B. $CH_3-CH_2-\underset{\underset{CH_3}{|}}{C}=CH-CH_3$

Solution

Write the IUPAC name for each of the following:

A. $CH_3-CH_2-C\equiv C-CH_3$ **2-pentyne**

B. $CH_3-CH_2-\underset{\underset{CH_3}{|}}{C}=CH-CH_3$ **3-methyl-2-pentene**

Cis and Trans Isomers

In an alkene, the double bond:

- is rigid
- holds attached groups in fixed positions
- makes cis/trans isomers possible

$$CH_3 \quad\quad CH_3$$
$$C = C$$
$$H \quad\quad\quad H$$
cis-2-butene

$$CH_3 \quad\quad H$$
$$C = C$$
$$H \quad\quad\quad CH_3$$
trans-2-butene

Cis-Trans Isomers

In *cis-trans* isomers:

- There is no rotation around the double bond in alkenes.
- Groups attached to the double bond are fixed relative to each other.

You can make a "double bond" with your fingers with both thumbs on the same side or opposite from each other.

Cis-Trans Isomers

Two isomers are possible when groups are attached to the double bond in different positions.

- In a **cis isomer**, groups are attached on the same side of the double bond.
- In the **trans isomer**, the groups are attached on opposite sides.

$$CH_3 \quad\quad CH_3$$
$$C = C$$
$$H \quad\quad\quad H$$
cis-2-butene

$$CH_3 \quad\quad H$$
$$C = C$$
$$H \quad\quad\quad CH_3$$
trans-2-butene

Cis-Trans Isomerism

- Alkenes cannot have cis-trans isomers if a carbon atom in the double bond is attached to *identical groups*.

Identical

$$H \quad\quad Br$$
$$C = C$$
$$H \quad\quad CH_3$$

2-bromopropene
(**not cis or trans**)

Identical

$$H \quad\quad Br$$
$$C = C$$
$$H \quad\quad Br$$

1,1-dibromoethene
(**not cis or trans**)

Cis-Trans Isomers in Nature

- Insects emit tiny quantities of pheromones, which are chemicals that send messages.
- The silkworm moth attracts other moths by emitting bombykol, which has one *cis* and one *trans* double bond.

$$H \quad H$$
$$H \diagdown C = C$$
$$C = C \diagdown CH_2CH_2CH_3$$
$$HOCH_2(CH_2)_7CH_2 \diagup \quad H$$

Naming *Cis-Trans* Isomers

The prefixes cis or trans are placed in front of the alkene name when there are *cis-trans* isomers.

← *cis* →

$$Br \diagdown \quad \diagup Br$$
$$C = C$$
$$H \diagup \quad \diagdown H$$

$$Br \diagdown \quad \diagup H$$
$$C = C$$
$$H \diagup \quad \diagdown Br$$

cis-1,2-dibromoethene   *trans*-1,2-dibromoethene

Problem

Name each, using *cis-trans* prefixes when needed.

A.
$$Br \quad Br$$
$$C = C$$
$$H \quad H$$

B.
$$CH_3 \quad H$$
$$C = C$$
$$H \quad CH_3$$

C.
$$CH_3 \quad Cl$$
$$C = C$$
$$H \quad Cl$$

Solution

A.
$$Br \quad Br$$
$$C = C$$
$$H \quad H$$
cis-1,2-dibromoethene

B.
$$CH_3 \quad H$$
$$C = C$$
$$H \quad CH_3$$
trans-2-butene

C.
$$CH_3 \quad Cl$$
$$C = C$$
$$H \quad Cl$$
1,1-dichloropropene
(Not *cis*- or *trans*-)

Alkenes: Compounds with C=C Double Bonds

- How many isomers are possible for a compound with the formula C_4H_8?

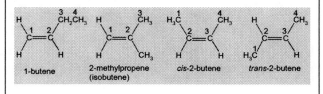

1-butene   2-methylpropene (isobutene)   cis-2-butene   trans-2-butene

Chapter 12

Gases and Their Properties

Chapter Goals

- Understand the basis of the gas laws and know how to use those laws (Boyle's law, Charles' law, Avogadro's hypothesis, Dalton's law).
- Use the (model) ideal gas law.
- Apply the gas laws to stoichiometric calculations.
- Understand kinetic-molecular theory as it is applied to gases, especially the distribution of molecular speeds (energies).
- Recognize why real gases do not behave like ideal gases.

Gases in Earth

Eleven elements are gases under normal conditions: five diatomic, H_2, N_2, O_2, F_2, Cl_2, and the six monatomic noble gases, He, Ne, Ar, Kr, Xe, Rn.

The thickness of the atmosphere is ~ 1/250 the diameter of the Earth. Yet this delicate layer is vital for our life. It shields us from harmful radiation and supplies substances needed for life, such as oxygen, nitrogen, carbon dioxide, and water.

Composition of the Atmosphere and Some Common Properties of Gases

Composition of Dry Air

Gas	% by Volume
N_2	78.09
O_2	20.94
Ar	0.93
CO_2	0.03
He, Ne, Kr, Xe	0.002
CH_4	0.00015
H_2	0.00005

Comparison of Solids, Liquids, and Gases

The density of gases is much less than that of solids or liquids.

Density (g/mL)	Solid	Liquid	Gas
H_2O	0.917	0.998	**0.000588**
CCl_4	1.70	1.59	**0.00503**

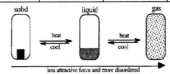

Gas molecules must be **very far apart** compared to liquids and solids.

General Properties of Gases

- There is a lot of "free" space in a gas.
- Gases can be expanded infinitely.
- Gases occupy containers uniformly and completely.
- Gases diffuse and mix rapidly.

Importance of Gases

- Airbags fill with N_2 gas in an accident.
- Gas is generated by the decomposition of sodium azide, NaN_3.

$$2\ NaN_3(s)\ \rightarrow\ 2\ Na(s)\ +\ 3\ N_2(g)$$

Properties of Gases

Gas properties can be **modeled** using math. Model depends on four quantities (parameters):

V = volume of the gas (L)
T = temperature (K)
n = amount (moles)
P = pressure
       (atmospheres)

Pressure

Pressure is force per unit area.

$$\text{Pressure} = \frac{\text{force}}{\text{area}} \qquad P = \frac{F}{A}$$

Force = F   N = Newton = $1\,kg{\cdot}m{\cdot}s^{-2}$

SI unit: pascal, $1\,Pa = \dfrac{N}{m^2} = 1\,kg{\cdot}m^{-1}{\cdot}s^{-2}$

Pressure

Atmospheric pressure (pressure of the atmosphere) is measured with a **barometer**, invented by **Torricelli** (1643).
Definitions of standard pressure
76 cm Hg = 760 mm Hg
= 760 **torr** = 1 atmosphere
= 1 atm
1 atm = 101.3 kPa = 1.013×10^5 Pa
1 bar = 1×10^5 Pa = 0.987 atm

H_2O density ~ 1 g/mL   Hg density = 13.6 g/mL

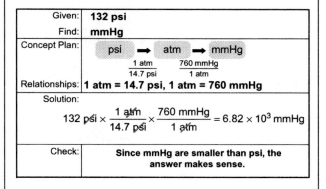

Pressure Unit Conversions
Now we can use pressure units as conversion factors.

Convert 202.6 kPa to Hg mm, bars, and atm.

$$202.6\ kPa \times \frac{760\ Hg\ mm}{101.3\ kPa} = 1{,}520\ Hg\ mm$$

$$202.6\ kPa \times \frac{10^3\ Pa}{1\ kPa} \times \frac{1\ bar}{10^5\ Pa} = 2.026\ bar$$

$$202.6\ kPa \times \frac{1\ atm}{101.3\ kPa} = 2.000\ atm$$

Example: A high-performance bicycle tire has a pressure of 132 psi. What is the pressure in mmHg?

Given:	**132 psi**
Find:	**mmHg**
Concept Plan:	psi → atm → mmHg $\frac{1\ atm}{14.7\ psi}$ $\frac{760\ mmHg}{1\ atm}$
Relationships:	**1 atm = 14.7 psi, 1 atm = 760 mmHg**
Solution:	$132\ psi \times \dfrac{1\ atm}{14.7\ psi} \times \dfrac{760\ mmHg}{1\ atm} = 6.82 \times 10^3\ mmHg$
Check:	**Since mmHg are smaller than psi, the answer makes sense.**

Manometers

- The pressure of a gas trapped in a container can be measured with an instrument called a **manometer.**
- Manometers are U-shaped tubes, partially filled with a liquid, connected to the gas sample on one side and open to the air on the other.
- A competition is established between the pressure of the atmosphere and the gas.
- The difference in the liquid levels is a measure of the difference in pressure between the gas and the atmosphere.

Manometer

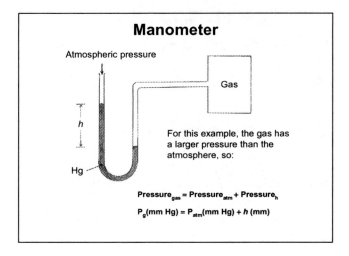

Atmospheric pressure

h

Hg

Gas

For this example, the gas has a larger pressure than the atmosphere, so:

Pressure$_{gas}$ = Pressure$_{atm}$ + Pressure$_h$

P$_g$(mm Hg) = P$_{atm}$(mm Hg) + h (mm)

IDEAL GAS LAW

$$P\,V = n\,R\,T$$

Brings together gas properties.

Can be derived from experiment and theory.

Boyle's Law   Robert Boyle (1627–1691)

If n(moles) and T are constant, then PV = k (constant).

This means, for example, that P goes up as V goes down.

A bicycle pump is a good example of Boyle's law. As the volume of the air trapped in the pump is reduced, its pressure goes up, and air is forced into the tire.

Boyle's Law:
The Volume-Pressure Relationship

$PV = k$

or $\quad V = k\,\dfrac{1}{P}$ or $V \propto 1/P$

$P_1V_1 = k_1$ for one sample of a gas.

$P_2V_2 = k_2$ for a second sample of a gas.

$k_1 = k_2$ for **the same sample of a gas (same number of moles at the same T).**

Thus, we can write Boyle's Law mathematically as: $\qquad P_1V_1 = P_2V_2$
for constant n and T

Boyle's Law: The Volume-Pressure Relationship

Example: At 25°C, a sample of He has a volume of 4.00×10^2 mL under a pressure of 7.60×10^2 torr. What volume would it occupy under a pressure of 2.00 atm at the same T?
Firstly, we need to convert atm to torr:

$2.00 \text{ atm} \times \dfrac{760 \text{ torr}}{1 \text{ atm}}$

$= 1520 \text{ torr}$

Then, using Boyle's law:

$P_1V_1 = P_2V_2$

$V_2 = \dfrac{P_1V_1}{P_2}$

$= \dfrac{(760 \text{ torr})(400 \text{ mL})}{1520 \text{ torr}}$

$= 2.00 \times 10^2 \text{ mL}$

Boyle's Law: The Volume-Pressure Relationship

- Notice that in Boyle's law, we can use any pressure or volume units as long as we consistently use the same units for both P_1 and P_2 or V_1 and V_2.

- Use your intuition to help you decide if the volume will go up or down as the pressure is changed, and vice versa.

Charles' Law Jacques Charles (1746–1823)

If n and P are constant, then
$V = kT$.

V and T are directly related (proportional).

$V \propto T$

Charles' Law: The Volume-Temperature Relationship; The Absolute Temperature Scale

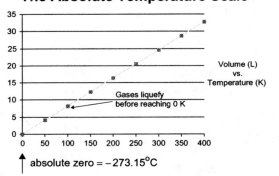

Volume (L) vs. Temperature (K)

Gases liquefy before reaching 0 K

absolute zero = −273.15°C

Charles' Law: The Volume-Temperature Relationship; The Absolute Temperature Scale

Charles' law states that the volume of a gas is directly proportional to the absolute temperature at constant pressure.

Gas laws must use the Kelvin scale to be correct.

Relationship between Kelvin and centigrade:

$$K = °C + 273.15$$

Charles' Law: The Volume-Temperature Relationship

Mathematical Form of Charles' law.

$$V \propto T \quad \text{or} \quad V = k\,T \quad \text{or} \quad \frac{V}{T} = k$$

$$\frac{V_1}{T_1} = k \quad \text{and} \quad \frac{V_2}{T_2} = k$$

$$\frac{V_1}{T_1} = \frac{V_2}{T_2} \quad \text{in the most useful form}$$

Charles' Law:

Example: A sample of hydrogen, H_2, occupies 1.00×10^2 mL at 25.0°C and 1.00 atm. What volume would it occupy at 50.0°C under the same pressure?

$$T_1 = 25 + 273 = 298 \qquad T_2 = 50 + 273 = 323$$

$$\frac{V_1}{T_1} = \frac{V_2}{T_2} \qquad V_2 = \frac{V_1 T_2}{T_1}$$

$$V_2 = \frac{1.00 \times 10^2 \text{ mL} \times 323 \text{ K}}{298 \text{ K}} = 108 \text{ mL}$$

Standard Temperature and Pressure

Standard temperature and pressure is given the symbol STP.

It is a reference point for some gas calculations.

Standard P $\equiv$ 1.00000 atm or 101.3 kPa

Standard T $\equiv$ 273.15 K or 0.00°C

Avogadro's Hypothesis

Avogadro's Hypothesis states that at the same T and P, **equal volumes** of two gases contain the **same number of molecules** (or moles) of gas. $V = k\,n$ n = moles

k = const.

On this basis, the following equation means:

$$2\,H_2(g) \quad + \quad O_2(g) \quad \rightarrow \quad 2\,H_2O(g)$$

2 moles H_2 reacts with 1 mole of O_2 to produce 2 moles H_2O

ratios **2 : 1 : 2**

2 L of H_2 reacts with 1 L of O_2 to produce 2 L of H_2O
   or

8 volumes + 4 volumes $\rightarrow$ 8 volumes (H_2O) $8 = 2 \times 4$

Here, volume means any unit (mL or L).

Gases in this experiment are measured at same T & P.

Avogadro's Hypothesis and the Standard Molar Volume

Avogadro's Hypothesis states that at the same temperature and pressure, equal volumes of two gases contain the same number of molecules (or moles) of gas.

If we set the temperature and pressure for any gas to be STP, then one mole of that gas has a volume called the **standard molar volume.**

* The standard molar volume is 22.4 L at STP.
   This is another way to measure moles.
   For **_gases_**, the volume is proportional to the number of moles. $V = k\,n$

11.2 L of a gas at STP = 0.500 mole
   44.8 L = ? moles

Avogadro's Hypothesis and the Standard Molar Volume

Example: One mole of a gas occupies 36.5 L, and its density is 1.36 g/L at a given temperature and pressure. (a) What is its molar mass? (b) What is its density at STP?

$$\frac{?g}{mol} = \frac{36.5\,L}{mol} \times \frac{1.36\,g}{L} = 49.6\ g/mol$$

$$\frac{?g}{L_{STP}} = \frac{49.6\,g}{mol} \times \frac{1\,mol}{22.4\,L} = 2.21\ g/L$$

The Combined Gas Law Equation

Boyle's and Charles' laws can be combined into one statement that is called the combined gas law equation.

Useful when the V, T, and P of a gas are changing, **but n is constant.**

Boyle's Law

$$P_1V_1 = P_2V_2$$

Charles' Law

$$\frac{V_1}{T_1} = \frac{V_2}{T_2}$$

For a given sample of gas:

$$\frac{PV}{T} = k$$

The combined gas law is:

$$\frac{P_1V_1}{T_1} = \frac{P_2V_2}{T_2}$$

The Combined Gas Law Equation

Example: A sample of nitrogen gas, N_2, occupies 7.50×10^2 mL at 75.0°C under a pressure of 8.10×10^2 torr. What volume would it occupy at STP?

$V_1 = 750$ mL $V_2 = ?$

$T_1 = 348$ K $T_2 = 273$ K

$P_1 = 810$ torr $P_2 = 760$ torr

Solve for $V_2 = \dfrac{P_1 \, V_1 \, T_2}{P_2 \, T_1}$

$= \dfrac{(810 \text{ torr}) (750 \text{ mL}) (273 \text{ K})}{(760 \text{ torr}) (348 \text{ K})}$

$= 627$ mL

The Combined Gas Law Equation

Example: A sample of methane, CH_4, occupies 2.60×10^2 mL at 32°C under a pressure of 0.500 atm. At what temperature would it occupy 5.00×10^2 mL under a pressure of 1.20×10^3 torr?

$V_1 = 260$ mL $V_2 = 500$ mL

$P_1 = 0.500$ atm $P_2 = 1200$ torr

$\quad = 380$ torr

$T_1 = 305$ K $T_2 = ?$

$T_2 = \dfrac{T_1 \, P_2 \, V_2}{P_1 \, V_1} = \dfrac{(305 \text{ K}) (1200 \text{ torr}) (500 \text{ mL})}{(380 \text{ torr}) (260 \text{ mL})}$

$= 1852 \text{ K} \approx 1580°C$

Summary of Gas Laws: The Ideal Gas Law

- Boyle's Law – $V \propto 1/P$ (at constant T & n)
- Charles' Law – $V \propto T$ (at constant P & n)
- Avogadro's Law – $V \propto n$ (at constant T & P)

Combine these three laws into one statement.

$$V \propto nT/P$$

Convert the proportionality into an equality.

$$V = \frac{nRT}{P}$$

This provides the **Ideal Gas Law** $PV = nRT$

R is a proportionality constant called the universal gas constant.

Summary of Gas Laws: The Ideal Gas Law

We must determine the value of **R**.
Recognize that for one mole of a gas at 1.00 atm, and 273 K (STP), the volume is 22.4 L.
Use these values in the ideal gas law.

$$R = \frac{PV}{nT} = \frac{(1.00 \text{ atm}) (22.4 \text{ L})}{(1.00 \text{ mol}) (273 \text{ K})}$$

$$= 0.0821 \frac{\text{L atm}}{\text{mol K}}$$

Summary of Gas Laws: The Ideal Gas Law

R has other values if the units are changed.
- $R = 8.314$ J/mol K
  Use this value in thermodynamics.

- $R = 8.314$ kg m^2/s^2 K mol
  Use this later in this chapter for gas velocities.

- $R = 8.314$ dm^3 kPa/K mol
  This is R in all metric units.

- $R = 1.987$ cal/K mol
  This the value of R in calories rather than J.

Using PV=nRT

What volume would 50.0 g of ethane, C_2H_6, occupy at 140ºC under a pressure of 1.82×10^3 torr?

$T = 140 + 273 = 413$ K

$$P = 1820 \text{ torr} \times \frac{1 \text{ atm}}{760 \text{ torr}} = 2.39 \text{ atm}$$

$$n = 50.0 \text{ g} \times \frac{1 \text{ mol}}{30.0 \text{ g}} = 1.67 \text{ mol}$$

$PV = nRT$

$$V = \frac{nRT}{P}$$

$$= \frac{(1.67 \text{ mol}) \left(0.0821 \frac{\text{L atm}}{\text{mol K}} \right) (413 \text{ K})}{2.39 \text{ atm}}$$

$$= 23.6 \text{ L}$$

Using PV=nRT

Calculate the pressure exerted by 50.0 g of ethane, C_2H_6, in a 25.0 L container at 25.0ºC.

$$MW = 30.07 \text{ g/mol}$$

$n = 1.67$ mol and $T = 298$ K

$$P = \frac{nRT}{V}$$

$$P = \frac{(1.67 \text{ mol}) \left(0.0821 \frac{\text{L atm}}{\text{mol K}} \right) (298 \text{ K})}{25.0 \text{ L}}$$

$$P = 1.63 \text{ atm}$$

A flying balloon contains 1.2×10^7 L of He at a pressure of 737 mm Hg and 25°C. What mass of He does the balloon contain? A.W. He = 4.00g/mol

$$737 \text{ mm Hg} \times \frac{1 \text{ atm}}{760 \text{ mm Hg}} = 0.970 \text{ atm}$$

$$PV = nRT \qquad PV = \frac{m}{M} RT$$

$$m = \frac{PVM}{RT} = \frac{0.970 \text{ atm} \times 1.2 \times 10^7 \text{ L} \times 4.00 \text{g/mol}}{0.0821 \text{ atm L/K mol} \times 298 \text{ K}}$$

$$m = 1.9 \times 10^6 \text{ g}$$

Determination of Molecular Weights and Molecular Formulas of Gaseous Substances

Example: A gaseous compound is 80.0% carbon and 20.0% hydrogen by mass. At STP, 546 mL of the gas has a mass of 0.732 g. What is its molecular formula?

100 g of compound contains 80 g of C and 20 g of H.

$$? \text{ mol C atoms} = 80.0 \text{ g C} \times \frac{1 \text{ mol C}}{12.0 \text{ g C}} = 6.67 \text{ mol C}$$

$$? \text{ mol H atoms} = 20.0 \text{ g H} \times \frac{1 \text{ mol H}}{1.01 \text{ g H}} = 19.8 \text{ mol H}$$

Determine the smallest whole number ratio.

$$\frac{19.8}{6.67} = 3 \therefore \text{ the empirical formula is } CH_3 \text{ with mass} = 15$$

Determination of Molecular Weights and Molecular Formulas of Gaseous Substances

Example: At STP, 546 mL of the gas has a mass of 0.732 g. **Now, we can calculate the molar mass (M) of the compound:**

$$PV = nRT, \quad n = \frac{m(g)}{M}, \quad PV = \frac{m}{M} RT, \quad M = \frac{m RT}{PV}$$

$$M = \frac{0.732 \text{ g} \times 0.0821 \text{ L.atm/mol.K} \times 273 \text{ K}}{1.00 \text{ atm} \times 0.546 \text{ L}} = 30.0 \frac{g}{mol}$$

$$\frac{M}{W EF} = \frac{30.0}{15.0} = 2 \quad \text{Then, } 2 \times CH_3 = \textbf{C}_2\textbf{H}_6 \text{, the formula.}$$

The Density of Gases and Molar Mass

As previously said, density of a gas is given in **g/L.** Calculate the density of methane, CH_4, at 37°C and a pressure of 1.50 atm.

$$PV = nRT, \qquad n = \frac{m(g)}{M}, \qquad PV = \frac{m RT}{M}$$

$$d = \frac{m}{V} = \frac{P M}{RT} \quad M(CH_4) = 12.01 + 4 \times 1.008 = 16.04 \text{ g/mol}$$
$$T = 37 + 273 = 310. \text{ K}$$

$$d = \frac{1.50 \text{ atm} \times 16.04 \text{ g/mol}}{0.0821 \text{ L.atm/mol.K} \times 310. \text{ K}} = 0.945 \frac{g}{L}$$

The Density of Gases and Molar Mass

The density of a gas is 0.391 g/L at 70.5 torr and 22.3°C. Calculate its molar mass (M).

$T = 22.3 + 273.15 = 295.4 \text{ K}$

$P = 70.5 \text{ torr} \times \dfrac{1 \text{ atm}}{760 \text{ torr}} = 0.0928 \text{ atm}$

$d = \dfrac{P M}{R T}$ Then, $M = \dfrac{d R T}{P}$

$M = \dfrac{0.391 \text{ g/L} \times 0.0821 \text{ L.atm/mol.K} \times 295.4 \text{K}}{0.0928 \text{ atm}} = 102 \ \dfrac{g}{mol}$

Mass-Volume Relationships in Reactions Involving Gases

- In this section, we are looking at reaction stoichiometry, like in Chapter 3, just including gases in the calculations.

$$2KClO_{3(s)} \xrightarrow{\ MnO_2 \ \& \ \Delta \ } 2KCl_{(s)} + 3\,O_{2(g)}$$

2 mol $KClO_3$	yields	2 mol KCL and 3 mol O_2
2 (122.6g)	yields	2 (74.6g) and 3 (32.0g)

Those 3 moles of O_2 can also be thought of as:
3(22.4L) or 67.2 L at STP

What volume of oxygen measured at STP can be produced by the thermal decomposition of 120.0 g of $KClO_3$?

$$FW \ (KClO_3) = 122.6 \text{ g/mol}$$

$$2\,KClO_{3(s)} \xrightarrow{\ MnO_2 \ \& \ \Delta \ } 2\,KCl_{(s)} + 3\,O_{2(g)}$$

$? \ L_{STP} \ O_2 = 120.0g \ KClO_3 \times \dfrac{1 mol KClO_3}{122.6g \ KClO_3} \times \dfrac{3 \ mol O_2}{2 mol KClO_3} \times \dfrac{22.4 \ L_{STP} \ O_2}{1 mol \ O_2}$

$? \ L_{STP} \ O_2 = 32.9 L_{STP} \ O_2$

---

31. Iron reacts with HCl(aq) to produce iron (II) chloride and hydrogen gas. The gas from the reaction of 2.2 g Fe with excess acid is collected in a 10.0 L flask at 25°C. What is the H_2 pressure?

$$Fe(s) + 2\,HCL(aq) \rightarrow FeCl_2(aq) + H_2(g)$$

$2.2 \text{ g Fe} \times \dfrac{1 \text{ mol Fe}}{55.85 \text{ g Fe}} \times \dfrac{1 \text{ mol } H_2}{1 \text{ mol Fe}} = 0.039 \text{ mol } H_2$

$PV = nRT$ $P = \dfrac{nRT}{V}$

$P = \dfrac{0.039 \text{ mol} \times 0.0821 \text{ atm L/K mol} \times 298 \text{ K}}{10.0 \text{ L}} = 0.095 \text{ atm}$

Gases Laws and Chemical Reactions

Bombardier beetle uses decomposition of hydrogen peroxide to defend itself.

$$2H_2O_2(liq) \rightarrow 2H_2O(g) + O_2(g)$$

Decompose 1.1 g of H_2O_2 in a flask with volume of 2.50 L. What is the pressure of O_2 at 127°C? Of H_2O? $M(H_2O_2) = 34.0$ g/mol

Firstly, calculate moles of H_2O_2 and O_2:

$$1.1 \text{ g } H_2O_2 \times \frac{1 \text{ mol } H_2O_2}{34.0 \text{ g } H_2O_2} \times \frac{1 \text{ mol } O_2}{2 \text{ mol } H_2O_2} = \textbf{0.016 mol } O_2$$

Gases Laws and Chemical Reactions

$$2H_2O_2(liq) \rightarrow 2H_2O(g) + O_2(g)$$

Now, with the moles of O_2 we calculate P of O_2:

$$P = \frac{n\,R\,T}{V} = \frac{0.016 \text{ mol} \times 0.0821 \times 400 \text{ K}}{2.50 \text{ L}} = \textbf{0.21 atm } O_2$$

We can calculate moles and P of H_2O or just, because the coefficient of H_2O is 2 and O_2's is one, moles and P of H_2O are the double of O_2's.

H_2O, n = 0.032 mol   P = 0.42 atm

Dalton's Law of Partial Pressures

For the reaction:
$$2H_2O_2(liq) \rightarrow 2H_2O(g) + O_2(g)$$
final pressures are: 0.42 atm   0.21 atm

They are said to be **partial pressures of H_2O and O_2.** What is the total pressure in the flask?

P_{total} in gas mixture $= P_A + P_B + \ldots$ Therefore,

$P_{total} = P(H_2O) + P(O_2) = 0.42 + 0.21 = 0.63$ atm

Example of **Dalton's Law**, $P_{total} = \Sigma$ (partial pressures)

holds at **constant volume and temperature. Each gas occupies the whole volume regardless of other**.

Dalton's Law of Partial Pressures

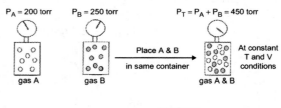

John Dalton
1766–1844

Dalton's Law of Partial Pressures

Vapor Pressure is the pressure exerted by a substance's vapor over the substance's liquid at equilibrium.

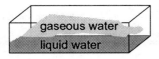

gaseous water

liquid water

Dalton's Law of Partial Pressures

Example: A sample of hydrogen was collected by displacement of water at 25.0°C. The atmospheric pressure was 748 torr. What pressure would the dry hydrogen exert in the same container?

$P_{total} = P(H_2) + P(H_2O)$ ∴ $P(H_2) = P_{total} - P(H_2O)$

$P(H_2O) = 24$ torr (**taken from tables**) at 25.0°C
(Appendix G, pp. A-20)

$P(H_2) = (748 - 24)$ torr $= 724$ torr

Dalton's Law of Partial Pressures

A sample of oxygen was collected by displacement of water. The oxygen occupied 742 mL at 27.0°C. The atmospheric (barometric) pressure was 753 torr. What volume would the dry oxygen occupy at STP? $P(H_2O) = 27$ torr (taken from tables) at 27.0°C

$V_1 = 742$ mL $V_2 = ?$

$T_1 = 300$ K $T_2 = 273$ K

$P_1 = (753 - 27) = 726$ torr $P_2 = 760$ torr

$V_2 = 742$ mL $\times \dfrac{273\ K}{300\ K} \times \dfrac{726\ torr}{760\ torr} = 645$ mL @ STP

Total P, Partial Pressures, and Mole Fraction

$P_A = 200$ torr $P_B = 250$ torr $P_T = P_A + P_B = 450$ torr

Place A & B in same container

At constant T and V conditions

gas A   gas B   gas A & B

For a mixture of several gases (A, B, C, …):

$P_T = P_A + P_B + P_C + ... = \dfrac{n_A R T}{V} + \dfrac{n_B R T}{V} + \dfrac{n_C R T}{V} + ...$

$P_T = \dfrac{(n_A + n_B + n_C) R T}{V}$ $P_i = \dfrac{n_i R T}{V}$ i = A, B, C,...

There is a simple relation between every P_i and P_T

Total P, Partial Pressures, and Mole Fraction

$$X_A = \frac{P_A}{P_T} = \frac{n_A \cancel{RT}}{(n_A + n_B + n_C + ..)\cancel{RT}} = \frac{n_A}{n_A + n_B + n_C + ..} = \frac{n_A}{n_T}$$

X_A is the **mole fraction of A**, ratio of moles of A to the total, or ratio of partial pressure to total pressure

$$X_A + X_B + X_C + ... = \frac{n_A}{n_T} + \frac{n_B}{n_T} + \frac{n_C}{n_T} +...$$

$$X_A + X_B + X_C + ... = \frac{n_A + n_B + n_C +...}{n_T} = 1 \quad \textbf{They add up to one.}$$

A 5.25 L sample of argon was collected over water at 30°C and at a pressure of 830.0 torr. What were the mole fractions of Ar and water?
Vapor pressure of H_2O is 31.8 torr at 30°C.

$P_T = P_{Ar} + P_W \therefore P_{Ar} = P_T - P_W \qquad$ W = water

$P_{Ar} = (830.0 - 31.8) = 798.2$ torr

$X_{Ar} = \dfrac{P_{Ar}}{P_T} = \dfrac{798.2 \text{ torr}}{830.0 \text{ torr}} = 0.9617$ **(no units)**

$X_{Ar} + X_W = 1 \therefore X_W = 1 - X_{Ar} = 1 - 0.9617 = 0.0383$

We didn't use V and T. We didn't need them.

Dalton's Law of Partial Pressures

If 1.00×10^2 mL of hydrogen, measured at 25.0°C and 3.00 atm pressure, and 1.00×10^2 mL of oxygen, measured at 25.0°C and 2.00 atm pressure, were forced into one of the containers at 25.0°C, what would be the pressure of the mixture of gases?
V and T of the two gases are the same: the Ps won't change

$$P_{Total} = P_{H_2} + P_{O_2}$$

$$= 3.00 \text{ atm} + 2.00 \text{ atm}$$

$$= 5.00 \text{ atm}$$

What is the mol fraction of each gas in the mixture?

$$X(H_2) = \frac{P(H_2)}{P_T} = \frac{3.00 \text{ atm}}{5.00 \text{ atm}} = 0.600 \quad X(O_2) = 1 - 0.600 = 0.400$$

KINETIC MOLECULAR THEORY (KMT)

Theory used to explain gas laws. KMT assumptions are:
- **Gases consist of molecules in constant, random motion.**
- **P arises from collisions with container walls.**
- **No attractive or repulsive forces between molecules. Collisions elastic.**
- **Volume of molecules is negligible, compared to the V of container.**

Kinetic Molecular Theory

Because we assume molecules are in motion, they have a kinetic energy.

$$KE = \frac{m\, u^2}{2} \qquad u = speed$$

At the same T, all gases have the same average KE.

$$KE = \frac{3}{2} RT \quad \text{(for 1 mol of gas)}$$

$$KE \propto T \text{ (KE is directly proportional to T)}$$

As T goes up, KE also increases — and so does speed.

_____

_____

_____

_____

Kinetic Molecular Theory

Maxwell's equation

$$U_{rms} = \sqrt{\overline{u^2}} = \sqrt{\frac{3RT}{M}}$$

root mean square speed

where u is the speed and M is the molar mass.

- speed INCREASES with T

- speed DECREASES with M

_____

_____

_____

_____

Kinetic Molecular Theory

Example: What is the root mean square speed of N_2 molecules at room T, 25.0°C?

$$u_{rms} = \sqrt{\frac{3\left(8.314\, \frac{kg\, m^2}{sec^2\, K\, mol}\right)(298\ K)}{0.028\ kg\,/\,mol}}$$

$$= 515\ m\,/\,s = 1159\ mi\,/\,hr$$

$$R = 8.314\, \frac{kg\, m^2}{s^2\, K\, mol}$$

M in kg

What is the root mean square velocity of He atoms at room T, 25.0°C?

$$u_{rms} = \sqrt{\frac{3\left(8.314\, \frac{kg\, m^2}{sec^2\, K\, mol}\right)(298\ K)}{0.004\ kg\,/\,mol}}$$

$$= 1363\ m\,/\,s = 3067\ mi\,/\,hr$$

_____

_____

_____

_____

Distribution of Gas Molecule Speeds

- **Boltzmann plots**

- Named for Ludwig Boltzmann, Scot physicist who also developed the theory of electromagnetic radiation (1864).

_____

_____

_____

_____

Velocity of Gas Molecules

Molecules of a given gas have a **range** of speeds.

Velocity of Gas Molecules

Average velocity decreases with increasing mass, at the same temperature:

$$u_{rms} = \sqrt{\frac{3RT}{M}}$$

GAS DIFFUSION AND EFFUSION

DIFFUSION is the gradual mixing of molecules of different gases.

Air   Br$_2$

GAS EFFUSION

EFFUSION is the movement of molecules through a small hole into an empty container.

GAS DIFFUSION AND EFFUSION

Molecules effuse thru holes in a rubber balloon, for example, at a rate (= moles/time) that is

• proportional to T

• inversely proportional to M.

Therefore, He effuses more rapidly than O_2 at same T.

GAS DIFFUSION AND EFFUSION

Graham's law governs effusion and diffusion of gas molecules.

Thomas Graham, 1805–1869 Professor in Glasgow and London

$$\frac{\text{Rate for A}}{\text{Rate for B}} = \sqrt{\frac{\text{M of B}}{\text{M of A}}}$$

Rate of effusion is inversely proportional to its molar mass.

GAS DIFFUSION AND EFFUSION

• Calculate the ratio of the rate of effusion of He to that of sulfur dioxide, SO_2, at the same: temperature and pressure.

$$\frac{R_{He}}{R_{SO_2}} = \sqrt{\frac{M_{SO_2}}{M_{He}}}$$

$$= \sqrt{\frac{64.1 \text{g / mol}}{4.0 \text{ g / mol}}}$$

$$= \sqrt{16} = 4 \therefore R_{He} = 4R_{SO_2}$$

GAS DIFFUSION AND EFFUSION

A sample of hydrogen, H_2, was found to effuse through a pinhole 5.2 times as rapidly as the same volume of unknown gas (at the same temperature and pressure). What is the molecular weight of the unknown gas?

$$\frac{R_{H_2}}{R_{unk}} = \sqrt{\frac{M_{unk}}{M_{H_2}}}$$

$$5.2 = \sqrt{\frac{M_{unk}}{2.0 \text{ g/mol}}}$$

$$27 = \frac{M_{unk}}{2.0 \text{ g/mol}}$$

$$M_{unk} = 27 \, (2.0 \text{ g/mol}) = 54 \text{ g/mol}$$

GAS DIFFUSION AND EFFUSION

Examples of relationships:
Gas A effuses (or diffuses) 5.2 times faster than gas B:

$$\frac{\text{rate A}}{\text{rate B}} = 5.2 = SQRT(\frac{M_B}{M_A})$$

Gas A effuses at half of the speed of gas B:

$$\frac{\text{rate A}}{\text{rate B}} = \frac{1}{2} = 0.5$$

Gas A takes triple the time it takes gas B to diffuse:
the more time it takes, the slower the gas is.

$$\frac{\text{rate A}}{\text{rate B}} = \frac{1}{3}$$

Gas Diffusion

relation of mass to rate of diffusion

• HCl and NH_3 diffuse from opposite ends of tube.

• Gases meet to form NH_4Cl

• HCl heavier than NH_3

• Therefore, NH_4Cl forms closer to HCl end of tube.

Problem

Each four tires of a car is filled with a different gas. Each tire has the same V, and each is filled to the same P, 3.00 atm, and 25.0°C. One contains 116.0 g air, another 80.7 g Ne, the third 16.0 g He, and the fourth has 160.0 g of an unknown gas.

a) All tires contain the same number of molecules, because they have the same volume, T, and P.

$$n = \frac{PV}{RT} \quad \text{\# of molecules are moles (n)} \times 6.02 \times 10^{23}$$

Problem

b) Due to the fact that all gases have the same number of molecules, the ratio of their masses corresponds to the ratio of the molar masses, because n(moles) is the same. The difference in the masses is due to their molar masses:

A.W. He = 4.0 g/mol

$$\frac{\text{molar mass unknown}}{\text{molar mass He}} = \frac{160.0 \text{ g}}{16.0} = 10.0 \text{ times heavier} \times 4.0 = 40.0$$

c) According to the KE = 3/2 RT, all gases have the same kinetic energy at the same temperature, but the average speed depends on their molar masses. According to:

$$u_{rms} = SQRT(\frac{3RT}{M})$$

the gas with the lowest M will have the highest speed, i.e., He.

Using KMT to Understand Gas Laws

Recall that KMT assumptions are:

• Gases consist of molecules in constant, random motion.

• P arises from collisions with container walls.

• No attractive or repulsive forces between molecules. Collisions elastic.

• Volume of molecules is negligible, compared to the V of container.

_____

_____

_____

_____

Avogadro's Hypothesis and Kinetic Molecular Theory

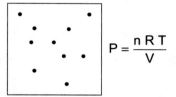

$$P = \frac{nRT}{V}$$

P proportional to n —
when V and T are constant

_____

_____

_____

_____

Avogadro's Hypothesis and Kinetic Molecular Theory

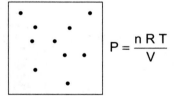

$$P = \frac{nRT}{V}$$

P proportional to T —
when n and V are constant

_____

_____

_____

_____

Avogadro's Hypothesis and Kinetic Molecular Theory

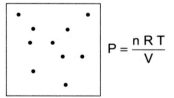

$$P = \frac{nRT}{V}$$

P proportional to 1/V —
when n and T are constant

_____

_____

_____

_____

Deviations from Ideal Gas Law

- Real molecules have **volume**.

- There are **intermolecular forces**.

  –Otherwise a gas could not become a liquid.

Deviations from Ideal Gas Law

Account for volume of molecules and intermolecular forces with **VAN DER WAALS' EQUATION.**

J. van der Waals, 1837–1923, Professor of Physics, Amsterdam. Nobel Prize 1910.

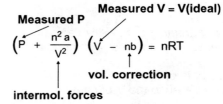

Measured P

Measured V = V(ideal)

$$\left(P + \frac{n^2 a}{V^2}\right)\left(V - nb\right) = nRT$$

vol. correction

intermol. forces

Deviations from Ideal Gas Law

Measured P   **Measured V = V(ideal)**

$$\left[P + \frac{n^2 a}{V^2}\right]\left(V - nb\right) = nRT$$

vol. Correction

intermol. forces

Cl$_2$ gas has **a** = 6.49, **b** = 0.0562
For 8.0 mol Cl$_2$ in a 4.0 L tank at 27°C,

$$P(\text{ideal}) = \frac{nRT}{V} = \frac{8.0 \times 0.0821 \times 300}{4.0} = 49.3 \text{ atm}$$

P (van der Waals) = 29.5 atm

Deviations from Ideal Gas Law

- Calculate the pressure exerted by 84.0 g of ammonia, NH$_3$, in a 5.00 L container at 200°C using the ideal gas law.

$$n = 84.0 \text{ g NH}_3 \times \frac{1 \text{mol}}{17.0 \text{ g}} = 4.94 \text{ mol}$$

$$P = \frac{nRT}{V} = \frac{(4.94 \text{ mol})\left(0.0821 \frac{\text{L atm}}{\text{mol K}}\right)(473 \text{ K})}{5.00 \text{ L}}$$

$$P = 38.4 \text{ atm}$$

Deviations from Ideal Gas Law

• Solve again using the van der Waal's equation.
• P = ? 84.0 g of NH_3, in a 5.00 L container at 200°C.

$n = 4.94$ mol $\quad a = 4.17 \dfrac{L^2 \, atm}{mol^2} \quad b = 0.0371 \dfrac{L}{mol}$

$$\left(P + \frac{n^2 a}{V^2}\right)(V - nb) = nRT \therefore$$

$$P = \frac{nRT}{V - nb} - \frac{n^2 a}{V^2}$$

$$P = \frac{(4.94 \text{ mol}) (0.0821 \frac{L \, atm}{mol \, K}) (473K)}{5.00 \text{ L} - (4.94 \text{ mol})(0.0371 \frac{L}{mol})} - \frac{(4.94 \text{ mol})^2 (4.17 \frac{L^2 \, atm}{mol^2})}{(5.00 \text{ L})^2}$$

$$P = \frac{191.8 \text{ L atm}}{4.817 \text{ L}} - 4.07 \text{ atm} = (39.8 \text{ atm} - 4.1 \text{ atm})$$

$P = 35.7$ atm which is a 7.6% difference from ideal